Mathematical Modelling Courses
for Engineering Education

NATO ASI Series

Advanced Science Institutes Series

A series presenting the results of activities sponsored by the NATO Science Committee, which aims at the dissemination of advanced scientific and technological knowledge, with a view to strengthening links between scientific communities.

The Series is published by an international board of publishers in conjunction with the NATO Scientific Affairs Division

A	Life Sciences	Plenum Publishing Corporation
B	Physics	London and New York
C	Mathematical and Physical Sciences	Kluwer Academic Publishers Dordrecht, Boston and London
D	Behavioural and Social Sciences	
E	Applied Sciences	
F	Computer and Systems Sciences	Springer-Verlag Berlin Heidelberg New York
G	Ecological Sciences	London Paris Tokyo Hong Kong
H	Cell Biology	Barcelona Budapest
I	Global Environmental Change	

NATO-PCO DATABASE

The electronic index to the NATO ASI Series provides full bibliographical references (with keywords and/or abstracts) to more than 30 000 contributions from international scientists published in all sections of the NATO ASI Series. Access to the NATO-PCO DATABASE compiled by the NATO Publication Coordination Office is possible in two ways:

- via online FILE 128 (NATO-PCO DATABASE) hosted by ESRIN, Via Galileo Galilei, I-00044 Frascati, Italy.

- via CD-ROM "NATO Science & Technology Disk" with user-friendly retrieval software in English, French and German (© WTV GmbH and DATAWARE Technologies Inc. 1992).

The CD-ROM can be ordered through any member of the Board of Publishers or through NATO-PCO, Overijse, Belgium.

Series F: Computer and Systems Sciences Vol. 132

Mathematical Modelling Courses for Engineering Education

Edited by

Yaşar Ersoy

Department of Science Education
Middle East Technical University
TR-06531 Ankara, Turkey

Alfredo O. Moscardini

School of Computing and Information Systems
University of Sunderland
Sunderland SR1 3SD, UK

Springer-Verlag
Berlin Heidelberg New York London Paris Tokyo
Hong Kong Barcelona Budapest
Published in cooperation with NATO Scientific Affairs Division

Proceedings of the NATO Advanced Research Workshop on The Design of Mathematical Modelling Courses for Engineering Education, held in Izmir, Turkey, July 12–16, 1993

CR Subject Classification (1991): G.1.8, I.1, I.6, J.2, K.3

ISBN 978-3-642-08194-1

CIP data applied for

Preface

As the role of the modern engineer is markedly different from that of even a decade ago, the theme of engineering mathematics education (EME) is an important one. The need for mathematical modelling (MM) courses and consideration of the educational impact of computer-based technology environments merit special attention. This book contains the proceeding of the NATO Advanced Research Workshop held on this theme in July 1993.

We have left the industrial age behind and have entered the information age. Computers and other emerging technologies are penetrating society in depth and gaining a strong influence in determining how in future society will be organised, while the rapid change of information requires a more qualified work force. This work force is vital to high technology and economic competitiveness in many industrialised countries throughout the world. Within this framework, the quality of EME has become an issue. It is expected that the content of mathematics courses taught in schools of engineering today have to be re-evaluated continuously with regard to computer-based technology and the needs of modern information society. The main aim of the workshop was to provide a forum for discussion between mathematicians, engineering scientists, mathematics educationalists, and courseware developers in the higher education sector and to focus on the issues and problems of the design of more relevant and appropriate MM courses for engineering education.

It is an observed fact that recent developments in the field of both hardware and software have provided most mathematics instructors in the industrialised countries with the opportunity to fundamentally rethink their teaching approach and an opportunity to explore the potential of computer-based environments (CBE) in teaching and learning. Computers and computer-based technology help to release engineers from the drudgery of data capture, calculating, analysing, and formatting the results. The emerging technology will make it possible for the future engineer to devote much more time to thinking, abstraction, and generalisation. In the last two decades, therefore, there has been a great effort to integrate computers into the mathematics curricula in various institutions and to support numerous projects for various uses of computers and their implementation in higher education. Among others, there is a general trend to inject computer algebra systems (CASs), e.g., *Derive, Maple, Mathematica,* etc., into the teaching of mathematics in the schools of engineering. We do hope that such trends will result in changes in how to teach as well as what to teach. Students can concentrate on understanding concepts in depth and be exposed to real world problems. However, very little is known about how individuals learn mathematics and what environments are most productive in stimulating this learning. After getting together and exchanging ideas between researchers and academics whose work is concerned with teaching engineering mathematics, it was hoped that an agreement on how to revitalise EME by means of the CBE would be reached in the workshop.

Among mathematics teachers, MM in the natural, applied, and social sciences is a favourite solution to the problem of how to use mathematics to solve real life problems. However, there is no general agreement on the most appropriate way to integrate MM in the existing courses in engineering curricula. Nevertheless, with the varied uses of computer and adequate software, it is now possible to provide new stimulating and relevant learning environments for either engineering science or MM courses, and prepare the student-engineers for the future, not the past. In this connection, the aims in designing MM courses might be to make the mathematics materials seem more relevant, and equip students so that they can recognise and cope with situations in natural and engineering sciences.

One hopes that this type workshop, which is a co-operation between the members of NATO countries and some non-NATO countries, will serve as a model to help the less developed member countries further their EME. Overall, we believe that the workshop and the follow-up publications will substantially enhance our understanding of the use of CASs in MM courses, and will help to improve the learning and teaching of engineering mathematics.

Alfredo O. Moscardini   Yaşar Ersoy
University of Sunderland   *Middle East Technical University*

June 1994

Acknowledgements

NATO has funded the workshop from which this book is an off-spring. We thank the NATO ARW Committee for their support and positive attitude toward the theme discussed in the workshop.

As always, many people contributed to the success of the workshop and to the potential success of this book. For his contribution for the organisation of the workshop, we thank Professor Fred Simons. For their computer skills and tireless work on the manuscript, we thank Petia Stoyanova and G. Alpay Ersoy.

Further, to all the participants and the ideas they shared during the conference and in this book, we are truly thankful.

Finally, most of the editing of this book was kindly supported by the METU during a part of Prof. Ersoy's sabbatical at the University of Sunderland, UK.

Contents

Introduction

The engineering discipline has seen many changes in the twentieth century. At the beginning of the century, it was neatly classified into many branches, e.g., thermodynamics, electromagnetic theory, fluid mechanics, heat transfer, mechanics, etc. In the last two decades, there is the beginning of a movement to recognise the common elements of these branches and thus to reunite the engineering field. These tendencies are apparent at university level by the number of universities that now run common first years to their engineering degrees, the introduction of mathematical modelling (MM) courses into the curricula, and the use of new computer technology and improved software. This book deals with MM and the use of new software in engineering education.

The term *mathematical modelling* has many meanings yet once a definition is put forward, most scientists and engineers claim that that is what they have always done. The whole of engineering could be described as a prime example of MM. Yet MM is neither engineering mathematics nor applied mathematics. Many people confuse it with applied mathematics but again this is not so. It is to do with applying mathematics. Emphasis should be placed on the word *applying* as opposed to the adjective *applied*.

Its importance as a discipline of study in its own right is still controversial and has only gained prominence over the last twenty years. Evidence for this is in the number of international conferences which have been held during this period and international journals which have been published. However, the teaching of MM in schools and universities is still a lively subject for debate in several countries and this issue is firmly addressed in Part One of this book.

MM can be described as the process of translating real world problems into mathematical problems and interpreting the mathematical solutions in terms of the original problem. Real world problems seldom translate exactly into mathematical problems, there is always a degree of mismatch or error, and mathematical solutions however elegant are not always practical solutions. As such it often necessary to idealise or simplify the problem by making assumptions. The choice and use of assumptions is a key element of MM. Also important is the idea of iteration, i.e., the solution is achieved by starting with a simple model and gradually refining it, usually by adding features which were assumed unimportant in the original simple first model. In this sense, MM falls naturally into the area of problem solving.

There are few aspects of our society that are not influenced in some way by the information revolution, especially involving the use of computers. The question is: *How will this change and influence the mathematical and engineering curricula?* The computer has now replaced the slide rule as a fundamental tool of a modern engineer. But the computer cannot handle the fundamental concepts. Engineers must still understand the underlying mathematics, physical principles, and assumptions inherent in the models before they can effectively use the new computer technologies. The effective use of advanced finite element software for stress analysis or non-Newtonian flow requires an understanding of non-linear behaviour in solids and liquids as well as an understanding of the principles of the finite element method, for example. The computer and its associated software enables more and quicker calculations to be made. Future engineers must master a broad range of computer skills as well as the fundamentals of engineering sciences in depth. How much traditional mathematics is now needed?

This question is asked because of the advances in computer technology and interactive software. The introduction of good computer software into engineering courses is essential. A recent example of this is the use of computer algebra packages in both mathematical and engineering courses. The advent of these packages can be compared with the advent of cheap pocket calculators fifteen years ago. Just as they had a tremendous effect on the teaching of arithmetic, the computer algebra packages will revolutionise the teaching of algebra and calculus.

This text is the result of a week-long NATO Advanced Workshop held in Cesme, Turkey in June 1993. The workshop chose two common engineering areas, Fluid Flow and Heat Transfer, and examined the ways of improving teaching methods. The proceedings fell naturally into four parts:

* ***Mathematical Modelling Issues***
* ***Engineering Mathematics Education***
* ***Mathematical Models in Fluid Flow and Heat Transfer***
* ***Workshops.***

The text is written in a style intended for engineers, computer scientists, mathematicians, and mathematical educators to encourage an interest in the design of more relevant mathematics courses for engineers. We are aware of the fact that it is essential to understand the fundamentals of engineering sciences and mathematical growth to see the full spectrum of the process of MM and of the potential of computer tools. Part One therefore begins with a paper that describes what is happening in Western industry today and sets a challenge to the educators. The educators respond with a paper on the design of MM courses and one on the introduction of computer algebra package, in particular, *Maple*, into the curricula. The section ends with the requirements for modelling courses in Fluid Flow.

Part Two consists of eight papers which concern *engineering education* in general. The first two deal with the introduction of modern software packages *Mathematica* and *Matlab* into mathematical and engineering courses at university level. These are followed by two papers on new approaches to the teaching of Finite Element Analysis for engineers. In contrast, the fifth paper addresses the finite difference approach to Computational Fluid Flow. The last two papers cover other software that could be useful to engineers.

Part Three takes the two major engineering areas addressed by this workshop, i.e., Fluid Flow and Heat Transfer, and gives details of two complete models that are used by practical engineers. Part Four is a short resume of the workshops.

In every chapter, the authors have been encouraged to impress their own personalities on their view of the phenomena, but this has been done within a framework of internal consultation and allocating time to revise the manuscript.

Y. Ersoy

PART ONE

Mathematical Modelling Issues

This part concentrates on issues involved with mathematical modelling. The first paper, by Professor Cross, defines what is happening today in Industry. His analysis is that mathematical modelling is firmly embedded into the infrastructure not only for large firms in the industrialised countries but also for small to medium ones. The major factor for this rise in importance is the recent development of good powerful hardware and user-friendly software. There is a need for someone called a mathematical modeller who would usually work in a team backed by software experts, mathematical experts and engineering experts.

The second paper picks up the challenge of how to produce such people. The author, Professor Moscardini, suggests several ways to design mathematical modelling courses. Some of these suggestions are highly individualistic, but the author encourages teachers and lecturers to develop their own styles as teaching modelling is very different to teaching engineering or mathematics.

The third paper takes up Professor Cross' other point about software and concentrates on a computer algebra package called *Maple*. Various experiences of using *Maple* in a mathematical modelling context are discussed in this section by Dr. Curran.

The section ends with a paper by Professor Ciray from Turkey who identifies the type of modelling required for teachers of Fluid Flow.

The Role and Practice of Mathematical Modelling in Industry Today

M. Cross

CNMPA, University of Greenwich, London, UK

Abstract. Increasingly, industrial organisations are embedding mathematical modelling into their operational infrastructure. The major factors in making mathematical modelling an accessible tool for industry are (i) the rapid development of computer technology, (ii) the emergence of powerful user environment software for geometric modelling, meshing and results visualisation, and (iii) the development of computational mechanics algorithms and codes. The paper addresses the importance and limitations of each of the above factors in enabling effective mathematical modelling in an industry context. Finally, a look forward is advanced with respect to what engineers who wish to exploit models will have to know.

Keywords. Modelling software, computational modelling, geometric modelling, industrial modelling, modelling tasks, multi-physics problems

1 Introduction

The growth of mathematical modelling as an industry tool over the past 20 years has been astounding. It would have been hard in 1970 to envisage the extent to which software products embedding mathematical models have been absorbed into the infrastructure of industry. Whilst it might be expected that the major players of the aerospace, defence and nuclear industries would be heavily involved in mathematical modelling, it is surprising the extent to which small organisations utilise modelling software at the heart of their business. Such organisations do not become involved in modelling because it *may* be useful; they cannot afford to! Yet, they focus 10-20 % of their investment budget on the computing environment and appropriate modelling software tools, then engage a high level computational scientist or engineer to run the tools in their context. Moreover, these engineers do not usually fit into their neat organisational structures and cause all kinds of other problems/demands.

Why then are small as well as large, well established companies beginning to build the use of modelling software into their core activities? It is simply because by using the models in the design process, problems can be identified technology. In fact, industrial mathematical modelling has always been a struggle between model functionality and the limitations of computer resources. The need to generate model predictions in a practical time-frame, constrains the level of sophistication of the modelling. Approximations are, employed therefore, and most models are a mixture of phenomenological and empirical equations.

Of course, the power of computer technology has increased many orders of magnitude over the past 30 years or so. Models that took 30 hours to run on mainframe computers in the early 1970's takes less than a minute on a PC486 today. Each time that processor power increases by an order of magnitude (every 3 years or so), it enables modellers to be more ambitious in their construction of models. Associated with processor speed great advances in visualisation technology have been made in the last decade. As a consequence not only have the modelling software tools become more sophisticated, so have the visualisation tools now used to interrogate simulation data bases.

The objective of this paper is to summarise the state-of-the-art in computational modelling today and indicate how it is likely to develop in the next few years.

2 The Industrial Mathematical Modelling Task

Increasingly, scientists and engineers in industry who are active in mathematical modelling, are really users of models rather than creators of them. Those who create the models either work in small highly specialist groups in large organisations, universities or, increasingly, engineering software houses. What the scientist or engineering conventionally does in industry is to apply modelling software in a particular context. In this sense, most of his activity is concerned with configuring a software tool to represent his process. Most industry scientists or engineers who use models are then limited to:

- *specifying the geometry, sub-domains and mesh of the problem;*
- *the material properties and initial condition of each sub-domain;*
- *the boundary conditions externally and, possibly, between sub-domains; and*
- *possibly, the time step for transient problems.*

Even in packaged software that limits the user to these tasks, the generation of reliable simulation results that can be used with confidence is not trivial. At every stage, a major task is model verification. The problem is, that every consistent set of equations with legal initial and boundary conditions is a model of something, it just may not be the model of your process. Those applying and the production route modified so that the manufacturing department can get it "right (almost) first time" and that the product itself can be guaranteed to be "fit for purpose". In other words, the manufacturers are beginning to use models as key tools to demonstrate to the client their approach to product quality. As such, increasingly, mathematical modelling is being perceived as important in the design, manufacturing and marketing sectors of organisations.

This is a long way from the concept of a mathematician working in a backroom trying to persuade someone, anyone, to take notice of what he has to say! What is it that has caused the move from the periphery to centre stage? The answer is simple - impact. As it is established in the different industry sectors that by using models, manufacturing problems can be avoided and designs optimised yielding clear financial benefits, this feeds back strongly to the boardrooms of the relevant organisations.

When organisations first became involved in modelling they will typically develop a group of scientists/engineers to focus on this area. But the activity is now so mature in many companies that the modelling group has been disbursed and absorbed by the functions they used to work for. In the same way that CAD is distributed throughout organisations, increasingly so are those whose expertise is modelling.

In this paper, some of the key issues that control the functionality of the modeller in industry will be addressed and the challenges for the next decade identified. It will be shown that these challenges are as much to do with the associated software technology as with model functionality.

3 Mathematical Modelling and Computers

Scientists have been developing and using mathematical models to understand the way the world (indeed, the universe) works for centuries. However, it was not inspiration of the great scientific triumphs that encouraged more prosaic scientists and engineers to use models, but something much more basic - the emergence of computer technology to automate the process of "crunching the numbers", that made modelling accessible to a wider audience. Indeed, it was the pioneering work of Richardson and Southwell, using rooms full of people as compute engines, that demonstrated the power of models to solve numerically, problems of vital importance to the engineering community.

It has always been true that the ability of models to represent reality, at least, in an engineering context, is limited by the power and functionality of the available computing resources. In a practical sense, it is not possible to consider the level of sophistication of models apart from the associated computer models in an industrial context have learnt the hard way, that all model results should be treated with circumspection until appropriate validation has been completed. Even then they should be treated with suspicion.

Those who actually develop models are increasingly using modelling software frameworks. In the area of fluid dynamics, CFD software such as PHOENICS and FLOW3D (to name but two) provide a reasonable degree of open access to the code so that very specific models can be developed by other than the code originators. These tasks involve much more than model configuration as outlined above, but involve the expression of a complex physical scenario and its implementation at a generic level in the modelling software framework. Such model developers are involved in both mathematical representations of the physics, chemistry and engineering constraints of the scenario they wish to represent, and with the numerical implementation of the model. The modelling software frameworks may relieve the modeller of the task of writing the modelling code from scratch; they do nothing to reduce the responsibility of the modeller to understand the algorithmic strategies employed by the software, and how this should be utilised to implement the model in an effective manner. It is true to say that the majority of modelling failures using such frameworks are because the model builder failed to have a sufficiently in-depth understanding of the numerical procedures involved and their limitations.

4 The Computer Technology Context

As indicated above, industrial mathematical modelling is inextricably linked to
the functionality and power of the computer technology. As such any speculation
as to how mathematical modelling will develop is predicated by the available
computer technology. Today most modelling groups base their activities around
colour workstations that deliver a processor performance in the range 5-15
Mflops. The graphics facilities in these systems typically from SUN, SGi, HP and
IBM, are good and will get much better in the next couple of years enabling rapid
rotations and walk throughs with virtually a real time response. In the last couple
of years a number of very sophisticated visualisation software tools have been developed. AVS, Explorer and Data Visualiser are typical of such tools. They are
highly modular tools which can process mesh based data from any source. They
are often programmable so that users can develop a series of filters to transform
the original data set in a myriad ways and produce output in a manner optimal for
a specific application. Of course, included in this context is the use of animation
and video as a viable form of output to represent transient behaviour.

Increasingly, the models used in industry involve three dimensional geometries.
In this context, the definition of the geometries, the boundary conditions, initial
conditions, material properties domains and the mesh for the numerical solution
all have to be addressed. This specification of a 3D model has been one of the
major headaches for modellers in the industrial context. They do not worry about
computer run times of 30 hours if it takes 2-3 months to set up a problem. Fortunately, there are now infrastructure software tools to help in this task. CAD environments such as PATRAN and IDEAs, and geometric modelling environments
such as FAM can minimise these problems. Of course, they enable surface
geometries to be imported directly from conventional CAD tools. Separate geometric sub-domains can be identified and both material properties and conditions
can be specified for each of these. Also, geometric area patches can be identified
and used to define both internal (between sub-domains) and external boundary
conditions.

Another major inhibitor in this context is the generation of the mesh. Computational modelling software tools generally employ either a structured (possibly
block structured) or an unstructured mesh. The construction of block structured
meshes is relatively straightforward, however, such meshes do not always provide
an accurate representation of physical geometry. The other route involves the use
of an unstructured mesh, typically with tetrahedral, wedge and/or hexahedral
shaped elements or cells. The generation of such meshes can be extremely complex. It is not that techniques have been developed; there are a number with De-
launey triangulation and advancing front as the main players, but such techniques
are simply not yet robust enough. For complex geometries, such as 3D shapes
with thin regions, these automatic mesh generators often fail. The most effective
approach at the time of writing is the medial distance method which requires a
degree of interaction with the user. This interaction is something of a drawback.
In reality, the only kind of interaction that should be required is a specification of
some measure of mesh density (or total number of nodes). Ideally, the mesh gen-

erator should distribute the mesh to best represent the impact of the geometry on
the calculations. Mesh generation is still a considerable inhibitor in the use of 3D
modelling in industry.

Currently the workstation market leaders are releasing machines which deliver
~100 Mflop performance. Moreover, with message passing tools, such as, PVM
and fast FDDI communication links, groups of these processors can be used as
parallel architectures. For example, a heat conduction code that solves a 500 x
500 mesh size problem in 727 secs on a single HP735 processor takes 115 secs on
a 7 processor system with an efficiency of 95%. In other words, the delivered performance from this parallel system assembled from standard components is giving almost 600 Mflops. Processor power will soon cease to be the limiting factor
in the accurate modelling of complex physical phenomena.

So in summary, what we have today are computing environments based upon:
- *workstation technology with UNIX, Motif and a host of other software environment standards;*
- *reasonably high performance processors, typically in the range 5-15 Mflop;*
- *reasonably high performance graphics facilities;*
- *generic and programmable mesh based data visualisation tools;*
- *geometrical modelling software tools that can enable reasonably straightforward specification of initial and boundary conditions within sub-domains of
  the geometry;*
- *user interactive mesh generation.*

What can we look towards in the next 5 or so years:
- *workstations with enhanced functionality on today but with all the software
  environment standards we are now familiar with;*
- *the processing performance of single processors will approach 200 Mflop
  and through the use of message passing formalisms (such as PVM) and very
  high speed/low latency inter-connections, many processors on one network
  will be able to be used in parallel to address a single problem (i.e. the
  practical limitation of computer power will be minimised);*
- *the processing performance of graphics facilities will increase and the
  visualisation tools will become more comprehensive;*
- *the user environments for the geometric modelling software will be refined
  and improved;*
- *mesh generation will become genuinely automatic.*

What then will the computational modellers do is the question? Let us consider
this issue below.

5 Computational Modelling Software Tools

Over the past 20 years or so, a wide range of modelling software tools have been
developed and marketed. Some have been very process specific, whilst others
have been generic within a specific context, and some of the latter have provided
the user with a great deal of freedom to implement their own models.

In the modelling of continuum phenomena, the solid mechanics sphere of interest has been dominated by the use of finite element (FE) methods, whilst the majority of fluid dynamics modelling has been carried out using finite or control volume (FV) techniques (*i.e.* derivatives of finite difference methods). Heat transfer and electromagnetic field problems have been solved effectively using both FE and VF techniques. In the last decade, many modelling ventures have involved coupled processes (*e.g.* thermal stress analysis, coupled fluid flow and heat transfer). However, such couplings have been restricted to those components of the physics that can be readily incorporated into a single solution framework. In the coming decade, though, there will be much more demand to address genuinely multi-physics processes.

An example of multi-physics is the traditional process of metals casting. Today in an appropriate industrial context this can involve the following continuum phenomena:

- *mould filling by the liquid metal involving complex free surface viscous (possibly, turbulent) flows;*
- *melt stirring by electromagnetic fields;*
- *cooling by conduction and residual convection;*
- *heat exchange by radiation;*
- *solidification and change of phase of the metal;*
- *deformation of the solidifying component and generation of a residual stress distribution.*

Of course, all these phenomena are coupled (when present) and so a comprehensive simulation is a significantly demanding algorithmic process.

Conventionally, FV algorithms have been associated with structured meshes. Whilst the use of structured meshes affords some simplicity in the formulation, data storage and matrix structures of FV procedures they are not constrained to be so. Indeed, in the past few years a number of FV commercial CFD codes implemented on unstructured meshes have been developed. In fact the essential difference between FE and FV formulations is the use of :

- *weighting functions equal to the shape functions for FE methods;* and
- *weighting functions equal to unity in the control volume and zero elsewhere for FV methods.*

This translates as solving throughout an element volume for FE methods and across element (or cell) faces for the FV methods. Other differences are associated more with tradition than being essential distinctions. For example, FE methods tend to assemble the whole matrix and solve for multiple right hand sides using direct solvers. Such an approach is effective for linear elliptic problems (such as elastic material loading and/or thermal conduction), but is increasingly inflexible as the problems become more non-linear (such as for Navier-Stokes flows). On the other hand, FV methods have used methods which enable greater interaction of the variables being solved at a point because their heritage in CFD (particularly multiphase) is extremely non-linear.

At Greenwich, we have been developing techniques and software tools to enable multi-physics problems to be addressed. The techniques are based upon finite value formulations and solution strategies, yet implemented in the context of an unstructured mesh. Here the challenges that have had to be addressed from the numerical procedures perspective were the generation of algorithms to solve:

- *the solid deformation in elastic and inelastic mode;*
- *electromagnetic fields coupled with the flow;*
- *free surface flows with simultaneous heat transfer and solidification.*

Routes to a solution of each of these issues have been identified and are now being pursued.

Once the procedures have matured they need to be embedded in an appropriate software framework. One has been under development at Greenwich for some time and is based on the recognition that:

- *the assembly of coefficients for the linear systems in the iterative procedures, is essentially based upon mesh geometry and material properties with a cell; as such, the code can be structured so that the coefficients can be generated automatically;*
- *boundary conditions can be made generic with a variety of linearisation procedures, and so automated;*
- *equation solvers can also be made generic and so included as standard options.*

The framework has been designed in FORTRAN77 with an open structure using object oriented programming paradigms. As such, it is straightforward to implement a wide variety of algorithms and solution procedures.

At the present stage, CVUM3D has the following facilities:

- *tetrahedral, wedge, hexahedral cell shapes fully tested and data structure facilities for general polyhedral cells;*
- *cell-centred, vertex-centred and mixed approximation regimes;*
- *a range of linear solvers including preconditioned conjugate gradient, SOR, Jacobi in either point-by-point or whole field mode;*
- *hooks into FEMGEN/FEMVIEW for pre- and post-processing.*

The following solution procedures have been implemented so far:

- *A simple based solution of the slightly compressible Navier-Stokes flow equations with all variables co-located at the cell centres, using a modified Rhie-Chow method to estimate velocities at cell faces, plus a number of differencing methods to specify the convection terms.*
- *A k-ε turbulence model has recently been added.*
- *An implementation of the enthalpy based solidification procedure of Chow and Cross which can be solved using both cell-centre and vertex centred regimes. It can also be coupled with the flow procedures to simulate solidification by convection and conduction.*
- *A full 3D elastic stress-strain solution procedure has been implemented and tested. It currently uses a vertex centred approach with the PCCG solver*

> *over the whole field when material properties are constant and*
> *point-by-point otherwise.*

It is certain that similar efforts are underway elsewhere in the computational
mechanics community.

6 The Skills Required by Industrial Modellers

For the vast majority of those involved in mathematical modelling in an industrial
context, the emphasis will be on applying generic models embedded in a software
tool to some specific problem. Since it is likely that the modelling software tools
will sit in the context of geometric modelling and visualisation environments,
then such modellers will need primarily:

- *expertise and familiarity with the contextual software tools to define the*
  problem geometry, sub-domains, material characteristics, initial and bound-
  ary conditions, the mesh and how to visualise the simulation results in an
  optimal fashion;
- *expertise in the actual problem to be modelled;*
- *a strategy for model application that embeds verification with real world*
  data at its heart;
- *a sufficient understanding of the underlying mathematical equations being*
  solved and the strategy employed in their solution, to operate the modelling
  software tools with some confidence.

It is obvious that these skills are as much to do with a facility to use the appropriate software technology, as they are with the details of the models, their implementation and the context of the problem.

For those engineers/scientists who will actually build models using frameworks
(such as PHOENICS or ABAQUS) then they simply cannot escape the need to
have an in-depth understanding of both the process physics and the algorithmic
procedures used in their solution. Naivety here has been the downfall of many a
project.

Clearly, parallel computing architectures provide the route to very high processor power for mathematical modelling. The key factor inhibiting its use by the
modelling community at the moment, is the difficulty in exploiting the parallel
technology. Although, basic message passing tools such as PVM are available, the
task of implementing a modelling software tool on a parallel system is essentially
a manual one. Considerable parallel computing expertise is required to effectively
map continuum physics modelling software onto such architectures and this is the
main inhibitory factor against the uptake of parallel technology at this stage. Fortunately, this problem is a transient one, for a number of sets of tools are currently
being developed which automates a large part of the task of parallelisation. Thus,
in the next few years it will be relatively straightforward to implement modelling
software tools on parallel architectures.

So what will be the challenges of the next decade for modellers? The answer is
simple - process physics. The efforts of the last decade have already served to

demonstrate the limitations of current physical understanding, particularly in multiphase processes. What will then be required is much greater strategic co-operation between modellers and laboratory experimentalists to tease out the phenomenological behaviour in a less empirical, more predictable fashion, if the problems of the year 2000 are to be addressed effectively.

Bibliography

1. Zienkiewicz, O. C., Taylor, R. L.: The Finite Element Method, Vols 1, 2. McGraw-Hill 1991
2. Patankar, S.V.: Numerical Heat Transfer and Fluid Flow. Hemisphere 1980
3. Minkowycz, W. J. *et al* (Eds): Handbook of Numerical Heat Transfer, Wiley 1988
4. Crank, J.: Free and Moving-Boundary Problems, Clarendon Press 1984
5. Zienkiewicz, O. C.: Computational Mechanics Today. Int J Num Methods Engng 34, 9-33 (1992)
6. Cross, M.: Computational Mechanics: A Software Technology Emerges. In B. J. Noye *et al* (eds) Computational Techniques and Applications (CTAC-91), pub AMS, 49-61, 1992
7. Cross, M.: Development of Novel Computational Techniques for the Next Generation of Software Tools for Casting Simulation. In: T. Piwonka *et al* (eds.) Modelling of Casting. Welding and Advanced Solidification Processes VI, TMS, 115-126, 1993.

The Teaching of Mathematical Modelling

A.O. Moscardini
School of CIS, University of Sunderland, UK

Abstract. This paper describes the problems concerning the teaching of mathematical modelling courses. It begins by attempting to set mathematical modelling in a broad philosophical context. It discusses general problem solving and different ways of thinking. A personal methodology that is used in Sunderland is discussed and the paper finishes with examples from both Heat Transfer and Fluid Flow.

Keywords. Mathematical modelling, problem solving, lateral thinking, Stella, systems dynamics

1 Introduction

This paper dovetails with the one presented by Professor Cross. In his paper he presented what is happening in Industry. He then left us with a challenge. How will we, as teachers, react. Not at PhD or postgraduate level but what mathematics should we teach to the undergraduates.

The last talk discussed teams working together in Industry. A typical team would consist of an engineer, a numerical analyst, a software engineer, an applied mathematician and a modeller. What does the modeller do? He must do something different to the others. This is a good question to think about and the answer helps decide the content of mathematical modelling courses.

I am by trade a mathematician. As such, when I look at the list of words in Table 1. I would tend to use the words in the left hand column to describe mathematics. Unfortunately, most new undergraduates would use the words in the right

Table 1 Two Descriptions

Beautiful	*Boring*
Seductive	*Old hat*
Exciting	*Repetitive*
Imaginative	*Dull*
Creative	*Uninteresting*
Boundless	*Closed*
Relevant	*Irrelevant*
Useful	*Useless*

hand column. This is the challenge we face: to interest such students in the beauty of mathematics. Then they will be motivated to progress to more complicated areas. Modelling courses can fulfill this function. I think therefore that they modelling courses themselves should not be subject specific. A typical student will take many such specific modules such as Computational Fluid Dynamics, Complex Analysis, Electrodynamics during his degree course. Besides these modules, there should be one or two called mathematical modelling. But, this begs the question: what exactly is mathematical modelling? Before we can answer that, we should first define what we mean by mathematics?

2 Mathematics and Mathematical Modelling

2.1 Definitions of Mathematics

The problem of defining mathematics was posed by the ancient Greeks and one of the first definitions belongs to the Platonic School [1]. Plato believed that mathematics exists outside of space and time. For example, somewhere, there exists a perfect right angled triangle. Every right angled triangle on Earth is an imperfect copy of this perfect one. The same can be said of all mathematical constructs. Mathematics, as we know it, is a shadow of a perfect form.

"God created the integers and man created the rest." This sentence was attributed to Kronecker and could be taken as a slogan of the constructionalist school [2]. They believe that everything must be constructed from the natural numbers. The university of the law of the excluded middle is denied and they do not recognise the validity of a non-constructionalist proof. This school continues under the aegis of Brouwer and his disciples but to me it is a cold bleak school and I do not think it is a good definition to set before students.

"Pure Mathematics is subject where no one knows what they are talking about or whether what they are talking about is true or false." This saying is attributed to Bertrand Russell and perhaps reflects his frustration of spending twenty years trying to tie mathematics down to logic and eventually being defeated. Russell concluded that it was impossible to derive from whence mathematics came and also decided that the notion of truth is a difficult one in mathematics [3].

"Mathematics consists of moving meaningless symbols around a piece of paper according to certain rules." So said perhaps the greatest mathematician of all time, David Hilbert. Mathematics was compared to a gigantic chess game. He took up Russell's challenge and said that one does not talk about truth or falsity in

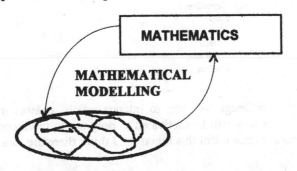

Fig. 1   A Symbolic Representation of Mathematical Modelling

"Mathematics consists of moving meaningless symbols around a piece of paper according to certain rules." So said perhaps the greatest mathematician of all time, David Hilbert. Mathematics was compared to a gigantic chess game. He took up Russell's challenge and said that one does not talk about truth or falsity in Mathematics but only consistency. This school is known as the Formalist School as it represents mathematics as all form and no meaning [4].

These are the four main schools and most mathematicians will believe in one of them. I suppose that the Platonic and Formalist Schools are most popular and both are similar in the fact that they are saying that *mathematics does not reside or exist in the real world.* It is separated from it. This is a most important position as it leads to the diagram shown in Fig. 1.

2.2 Definition of Mathematical Modelling

Fig. 1 shows a diagram of the world and mathematics is represented as a self consistent body of knowledge, divorced from the world, something like an inter-galactic cloud. When man has a problem he reaches up into the cloud and selects some appropriate mathematical symbols or equations *i.e.* builds a mathematical model. He then eventually transfers the mathematical solution back to the real problem. This *process* of entering and leaving the self consistent mathematical cloud is what I define as mathematical modelling. It is important to realise that there will never be a perfect fit. between the mathematics and the real world problem. There will always be errors. in matching. It is important to stress this point to students. Notice also that I define mathematical modelling as a process. A Mathematical Modelling course is a different course to one entitled Mathematical Models.

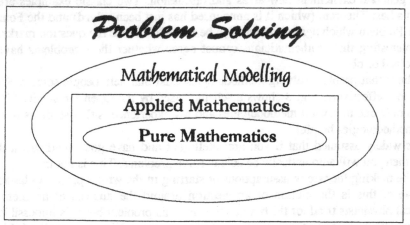

Fig. 2 Problem Solving and Mathematics

Fig. 2 expands the point even more. This diagram is maybe rather contro-versial but it represents my conception of various relationships. Whether you agree with this relationship or not, it is important for teachers to have their own conceptualisations of such things and to share them with the students. Mathematical Modelling courses are personal courses. They should present opportunities for discussion. They should establish an intellectual environment where there are no right and wrong answers only opinions. Students should realise that mathematics is not a dry, arid subject but a live one that people can actually argue about. Students never dream that there is scope for arguments in mathematics.

From Fig. 2 we see that mathematical modelling is part of problem solving. I would now like to talk about this subject.

3 Problem Solving and Thinking

3.1 Problem Solving

Problem solving involves thinking. Are students asked to think at University? Most courses demand that students learn techniques. Very rarely do they have to think. Even mathematics courses often just examine established models. The students aren't asked to think about how the equations were derived. Mathematical modelling courses should provide opportunities for students to think.

Problems can be described as *"open"* or *"closed"*. *"Closed"* problems are ones that have a unique solution that is normally obtained by a unique method. We tend to teach closed problems in schools and universities. The reasons are obvious. There are easier to teach, they are easier to assess and they provoke little discussion. *"Open"* problems are much more interesting. They have many or maybe no solutions, it is not obvious which mathematical method to use and quite often as not generate excitement new areas and discussion. Two classic examples are Fermat's Last Theorem (which it is announced has just been solved) and the Four Colour Problem which again is claimed to be solved?? Notice the question marks. It is interesting that mathematicians cannot agree whether these problems have been solved or not.

I believe that problem solving methodologies exist that help people solve problems more effectively. Fig. 3 shows a scheme used by the Open University [5]. This is only one of several methodologies and my experience is that students find such methodologies helpful.

It is widely assumed that if you are intelligent and have good mathematical bility, then you will automatically be able to solve problems. This is not true. You might be making the wrong assumptions or starting in the wrong place. A classic example of this is the motion of an electron around the nucleus of an atom. Classical physicists tried for thirty years to solve this problem but it is impossible to solve within the boundaries of classical physics. Once Nils Bohr realised this and decided to abandon classical physics then Quantum Physics was born and the problem soon solved. This example illustrates the role of thinking in problem solving. One can compare thinking to digging a hole; the harder one thinks, the deeper the hole. One is digging to reach the solution. But, if you are in the wrong hole then you may dig as deep as you can but you will never find a solution which may be just below the surface elsewhere. "Experts" are always inclined to stay in their own hole as this is one they themselves have constructed but truly creative people have the ability to find the right hole at the right time.

Problem Solving

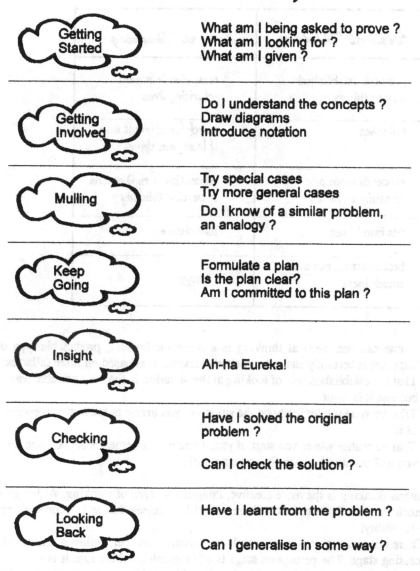

Getting Started	What am I being asked to prove ? What am I looking for ? What am I given ?
Getting Involved	Do I understand the concepts ? Draw diagrams Introduce notation
Mulling	Try special cases Try more general cases Do I know of a similar problem, an analogy ?
Keep Going	Formulate a plan Is the plan clear? Am I committed to this plan ?
Insight	Ah-ha Eureka!
Checking	Have I solved the original problem ? Can I check the solution ?
Looking Back	Have I learnt from the problem ? Can I generalise in some way ?

Fig. 3 Stages in Problem Solving

3.2 Thinking

Two types of thinking are defined by De Bono [6]: (i)Vertical thinking, and (ii) Lateral thinking. Their differences are shown in Table 2.

Table 2. Different Types of Thinking

Vertical Thinking	*Lateral Thinking*
preserve established way of things	recognises dominant polarising ideas
polarises	search for different way of looking at things
proceeds from one certainty to next	relaxation of rigid control of vertical thinking
fits into boxes	uses chance
breeds arrogance and smudginess	*PO*

As one can see, vertical thinking is a more methodical, predictable type of thinking that is certainly easier to teach and assess. It is based on three fallacies:
- That the established way of looking at the situation is the only possible way because it is right.
- That by working logically on the situation, you arrive at the best perception of it.
- That no matter where you start, if your logic or mathematics is good enough you will eventually reach the correct solution.

Lateral thinking is the more creative, imaginative type of thinking. Although it is much more difficult to teach, there are still exercises that will improves ones natural ability.

There are two major stages in problem solving:- the perception stage and the processing stage. The perception stage is often overlooked because it is often obvious where to start especially with closed problems. This is the stage where lateral thinking is important. Once one has started and has some idea of what to do, the processing stage commences. This is where vertical thinking and its logical methodicalness is important.

4 Modelling Methodology

We now can discuss mathematical modelling. Although, modelling is not a linear process and is rather unstructured, students must start somewhere and so I teach them an initial methodology. I don't intent that this becomes a rigid practice and that the students should be made to follow one and only one methodology. They hopefully will develop their own in time. The methodology taught at Sunderland is summarised by the diagram in Fig. 4.

PROBLEM ANALYSIS	MODEL IDENTIFICATION	IMPLEMENTATION
Reception	Modelling	Solution*
Formulation	Representation*	Interpretation
Conception	Analysis*	Validation*

Fig. 4 Modelling Methodology

As one can see, there are three major stages each subdivided into three sub-stages. This diagram is interesting in several ways. One sees that one third of the process is analysing the problem. If one notices the starred sections, these denote the mathematical sub-stages. This is surprising for two reasons:- the mathematics does not start until the process is half way through and that it itself is only one third of the total process. Many mathematicians don't like this. If so. let them invent their own methodology.

The Problem Analysis stage is a badly neglected stage. Much more research needs to be done here. This stage corresponds roughly to the perception stage mentioned previously. It is shown in greater detail in Fig. 5. Notice the arrows at the bottom of the diagram. These indicate the iterative nature of the process. Some of the major questions and processes are listed. The aim at the end of this stage is that the student has understood the problem, constructed an aim and a set of objectives and via a brainstorming session produced a list of possible features.

Fig. 6 covers the important stage of assumptions. The decision of whether to include a particular feature or not is made on the basis that tone first builds a simple model. As long as the discarded features are listed then the can be inserted into the second, third or fourth iteration. Getting the students to keep it simple is one of the most difficult parts of the whole procedure. Building the model is also very difficult and an aid to this is discussed in Section 5. The immediate reaction of most students after they have produced their model is to solve it. The Analysis stage attempts to make the student hold back and first analyse what he has produced. Only when he is certain that he has reasonable equations is it worth the effort to solve them.

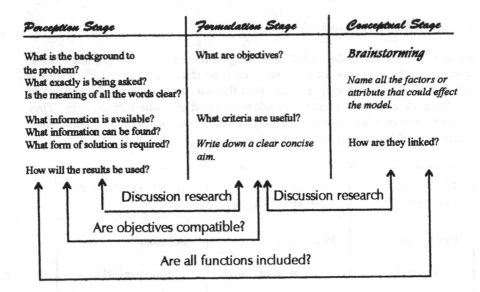

Fig. 5 Problem Analysis Stage

Fig. 6 Model Identification Stage

Fig. 7 shows the third major stage. Good software exists to help in the solution stage. It is important to then get the student to criticise the solution. There will then be several more iterations of the process before reasonable results are produced.

All through the process the emphasis is on thinking and criticising rather than elegant mathematical solutions. I tend to concentrate on the earlier model building stages and use software to cover the solution stages. These techniques are normally covered by Applied Mathematics courses.

Solution Stage	Interpretation Stage	Validation Stage
Appropriate choice of specific technique	Interpretaion of solution in the context of original problem	How valid is the model?
Test of stability of any numerical method	Any sensitivity tests?	How good are the predictions it makes?

Fig. 7 Implementation Stage

What are the problems of teaching these courses. The biggest problem is to teach students how to formulate their own equations. They find this exceedingly difficult. A good choice of suitable examples helps. These should be graduated and build up the difficulties. Some people worry about the students mathematical ability. I have never found that a problem. I tend to teach my course to the mathematical ability of the students. If they have limited ability then their models will be rather simple. But this is not important at this stage. The emphasis is on the process of building their own models and questioning them.

5 Examples

Nowadays, students are helped by very good software that is available on PCs and MACs. Some names are *Stella, Mathematica, Maple, Matlab, Extend*. Some of these will be discussed in detail later. I would like to discuss the use of *Stella* which is available on the Apple Macintosh machines. This software is useful because to me, it and its associated methodology is a perfect intermediary stage between the problem and mathematics. It therefore fits very nicely into the Modelling sub stage of Fig. 4.

This conference is concerned particularly with Heat Transfer and Fluid Flow, so I have decided to take two examples in these areas to illustrate some points. Both examples were produced by a colleague at the University of Sunderland, Mr Don Prior, who has been a staunch modelling ally for many years but unfortunately cannot be present at the workshop.

5.1 Example One: A Heat Transfer Problem

Fuel is fed to a burner A which is switched on and off by the thermostat B. The burning fuel gases give up some of their heat to the water of the central heating system in the heat exchanger compartment C and lose the rest of their heat to the atmosphere through the flue D. The water is pumped from the exchanger to the radiators in the room. During its journey, the water loses some heat to the sur-roundings outside the room and to the under-floorboard space. The radiators raise the room temperature and this is read by the thermostat B which switches the boiler burner on or off accordingly. The room loses heat to the outside world through its walls, doors and windows. A pictorial representation of the problem is shown in Fig. 8.

This is a realistic problem yet it involves many of the basic concepts of heat transfer: radiation, convection, conduction and all their related problems. The students would discuss this problem as per the problem analysis stage mentioned

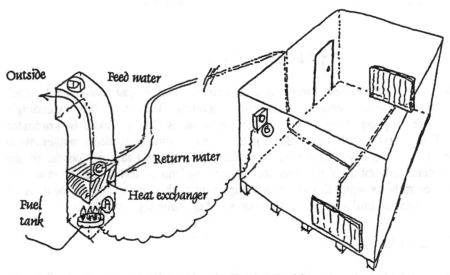

Fig. 8 Central Heating Problem

previously. They would brainstorm and make assumptions but then they come to the stage of building the model. Here I introduce System Dynamics.

This is a methodology that is based on feedback loops. The first stage is to build a Causal Model such as the one shown in Fig. 9. This is a typical first attempt at a causal loop diagram. It is reasonably easy to follow and we can see that the problem consists of a number of negative feedback loops that are controlling the process. Most students can reach such a diagram fairly easily.

This needs more detail and Fig. 11 overleaf shows a more complicated causal loop. One interesting thing in this diagram is the presence of the k-parameters. What are these parameters. What are their units. The student is now beginning to delve into the intricacies of the theory.

There is a very good piece of software called *Stella* which takes a causal loop model and then very easily transfers it into a *Stella* model which can be simulated.

Some results of the simulation are shown in Fig. 11 and one can see that the thermostat is working and controlling the temperature.

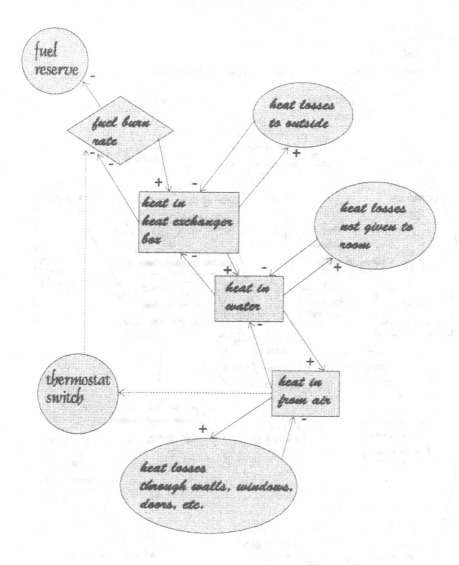

Fig. 9 Influence Diagram

What has the student learnt from this exercise. He has learned to produce a simple model and thus solve a problem. He has also learnt something about heat transfer. In order to get sensible results, the meaning of the k-parameters in Fig. 10 must have been discovered and the processes involved in radiation understood. The parameters are a key issue. In mathematics they are often introduced to balance the dimensions of the equations. Here they have real meaning and are essential to the simulation. One will also notice the delays in Fig. 10. These are difficult to handle mathematically but easy to introduce in *Stella*. They provide a much more realistic model and are a key element in this particular example.

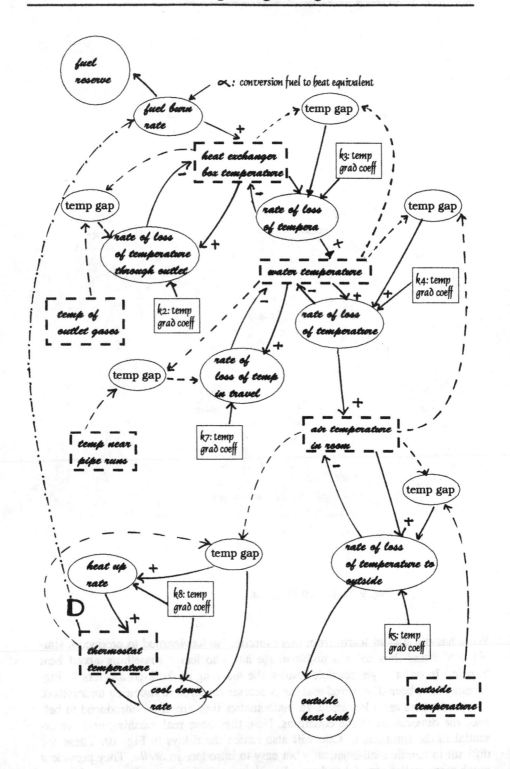

Fig. 10 Causal Loop Model

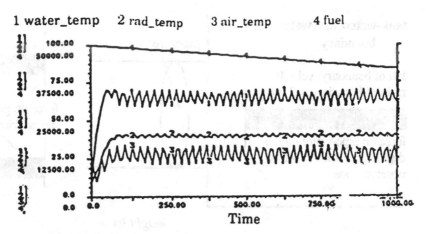

Fig. 11 *Stella* Output

After building a model, such as this, when the student has to study a "normal" mathematical module on Heat Transfer, there is more chance that he will relate to it, recognise behaviour patterns and generally better understand what is going on. I think that such an example is a useful first stage in any heat transfer course.

5.2 Example Two: A Fluid Flow Problem

I will now take a similar example in Fluid Flow. Here, we have what a student might think is the simplest possible situation *i.e.* water flowing through a pipe out of a container. In fact this is very difficult to model mathematically. The normal approach is to produce some very complicated generic partial differential equations and then simplify them. There is a particular approximation to the solution called Torricelli's law which covers this situation and is always quoted. Denied any reference books, most mathematicians would take some time to derive it from scratch. We ask the students to produce a system dynamics model. The student will need more understanding of physics in this example than the last. Terms such as impulse, momentum, acceleration, friction are needed. The diagram shown in Fig. 12 is a typical causal loop model.

Fig. 13 shows the *Stella* model and Fig. 14 shows a plot of the water from the pipe. The model result called plug-velocity is compared with Torricelli's formula and one can see that there is good agreement except at the end of the simulation. This is when the water has drained from the tank. Obviously, the velocity must then be zero but the model still produces a non-zero value. Obviously something is not working correctly. Can you see what it is?

A student who can build this model is beginning to obtain real insight in Fluid Flow. Again my previous comments as to what has been learnt are relevant here.

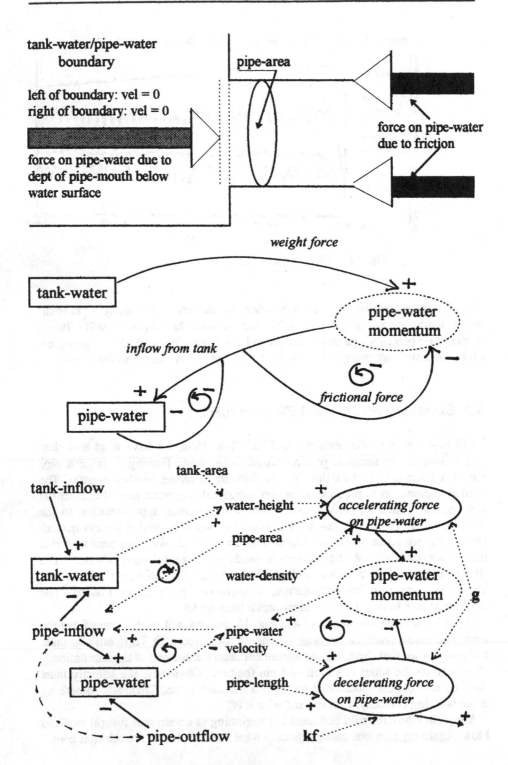

Fig. 12 Fluid Flow Problem

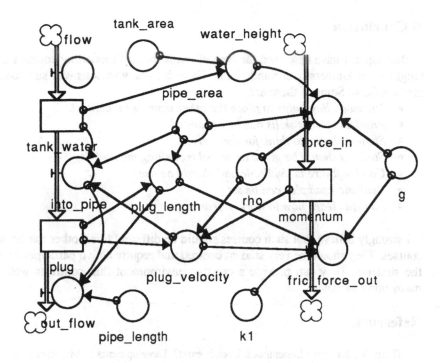

Fig. 13 *Stella* Model

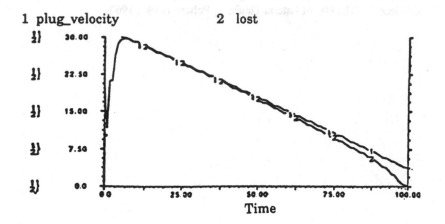

Fig. 14 *Stella* Output

6 Conclusion

In this paper, I have described the type of mathematical modelling course that is taught at the University of Sunderland. There are many reasons why such courses are beneficial. Some of these are:

- *The modelling helps to place the other courses in context.*
- *Problem solving skills can be enhanced*
- *Students begin to think for themselves*
- *Students should be given an initial modelling methodology*
- *Good software is available and should be used*
- *Realistic examples can be set.*
- *Students learn how to work in teams.*

I strongly believe that such courses should be different to all other mathematics courses. They should be very student centred and require active participation from the students. They can provide a creative environment that contrasts well with many other modules.

References

1. Thom, R.: Modern Mathematics: Does it exist? Developments in Mathematical Education, Cambridge University Press, 1973
2. Dieudonne, J.: Modern Axiomatic Methods and the Foundations of Mathematics. Great Currents of Mathematical Thought, Vol. 2, 251-266. New York: Dover 1971
3. Russell, B.: The Principles of Mathematics. Cambridge University Press, 1934
4. Curry, H.B.: Outlines of a Formalist Philosophy of Mathematics. Amsterdam: North Holland, 1951
5. Mathematical Foundations. M101 Open University, 1989
6. De Bono, E.: The Use of Lateral Thinking, Pelican Books, 1967.

Using Maple in a Mathematical Modelling Course

D.A.S. Curran and W. Middleton
Department of CIS, Sunderland University, UK

Abstract. The authors have been involved in the use of computer algebra systems in mathematics teaching for the past four years. In particular they have together with other colleagues received substantial government funding for a project involving the use of a computer algebra system (*Mathematica*) in designing an introductory, computer based, modelling course. The authors have also been involved in a TEMPUS project aimed at introducing new approaches to the teaching of mathematics to engineers at the Technical University of Sofia. This has involved the use of *Maple* in teaching mathematical modelling to graduate students. More recently they have used a totally computer algebra (*Derive/Maple*) based approach to teaching first and second year calculus and algebra courses. The authors have also used computer algebra to support the teaching of a course in dynamical systems on two MSc programmes. This paper outlines the authors' experiences in these areas and, in particular, details how *Maple* can be used to support and enhance the teaching of a dynamical systems course.

Keywords. Maple, modelling projects, dynamic systems, phase space, logistic mapping, NODE package

1 Introduction

Computer algebra packages have undergone considerable development over the last few years. At present the main contenders for the computer algebra crown are *Maple*, produced by the University of Waterloo and *Mathematica*, produced by Wolfram Research. They are both available on the typical platforms to be found in a university mathematics department, PCs, Macs, Sun workstations; they both have the same symbolic, numerical and graphical capabilities; they both have an educational pricing policy including special student versions. *Mathematica* has a more sophisticated programming language including elements of functional programming. *Maple*'s language is designed in the style of traditional procedural languages such as *Basic* and *Pascal*.

The advent of these packages has important consequences for the teaching and assessing of university mathematics courses. At present there is no generally agreed view on the optimum way of incorporating these packages into mathematics programmes. There is a whole spectrum of opinions ranging from completely replacing traditional mathematics teaching with a computer based approach using the packages to the view that the packages are only safe in the hands of an already professionally qualified mathematician. A number of universities in the UK are already using computer algebra packages in their undergraduate and postgraduate teaching. For both financial and hardware reasons the most commonly used package is *Derive*, with *Maple* second and only a handful, to the authors' knowledge, using *Mathematica*. *Derive* is the least powerful package and is only available on PCs. *Derive* will run within the 640K DOS limit without a maths co-processor. In contrast current PC versions of *Mathematica* and to a lesser extent *Maple* consume vast quantities of extended memory and require a maths co-processor to run at acceptable speeds. User groups for *Derive* and *Maple* have been set up under the auspices of the *Computers in Teaching Mathematics Initiative* based at Birmingham University. Members of these groups have regular meetings and exchange worksheets and other teaching materials.

2 Computer Algebra at Sunderland University

Staff at Sunderland University have been involved with computer algebra for a number of years. The first involvement was in the use of *Mathematica* in a project undertaken for TEED. This project was concerned with the design and implementation of a computer based mathematical modelling course. First year engineering students at Sunderland have been given a short *Derive* course for the last three years. This course covers using *Derive* for basic calculus and algebra. Although the course is a formal part of the curriculum the students are not assessed. For the past three years *Maple* has been used in teaching a number of modules in Master's programmes in mathematical modelling. The package has been used in courses on non-linear optimisation and dynamical systems. The University has an ongoing EC funded TEMPUS project with the Technical University of Sofia. This project involves the use of new approaches to mathematics teaching. As part of this staff at Sunderland have delivered a mathematical modelling course in Sofia to postgraduate students in which *Maple* was used as a tool to analyse Leslie matrix models of population growth. A more recent and radical departure has been the use of computer algebra in teaching and assessing first and second year calculus and algebra modules. These courses were completely computer based with all material being presented using computer algebra. Two packages were used. *Derive* for first year students and *Maple* for both first and second year.

2.1 Computer Algebra and Mathematics Teaching

At present a wide range of opinions exist about how computer algebra can be best incorporated into a mathematics teaching programme. Possible roles for computer algebra are

- As a mathematical assistant;
- As a means of carrying out mathematical experiments;
- As a living mathematical textbook;
- As a mathematical programming language;
- As a backup to traditional lectures and tutorials;
- As a replacement for traditional teaching on an 'appropriate' part of the curriculum;
- As a complete replacement for traditional teaching.

Within these roles computer algebra packages can be employed in various ways

- As a numerical solver;
- As a graphing tool;
- As a tool for carrying out tedious mathematical manipulations;
- As a workbench for experimental mathematics.

If courses are to be taught with the aid of a computer algebra package it is of course essential that they are also involved in the assessment. Assessment using computer algebra may take a number of forms

- Long term, unsupervised assignments of an exploratory nature;
- Shorter, time-constrained, supervised assignments;
- Mathematical experiments;
- Time constrained examinations.

Of course there may be many other roles, uses and assessment methods. One major problem in teaching courses using computer algebra is that it is new and demands a new approach to course delivery, student based tasks, assessment and all the other ingredients of the learning process. At present very few academic staff and even fewer students have any experience of the learning process using computer algebra. There are few relevant textbooks and almost none with suitable examples. Assessment experience, at least within the UK, is practically non-existent. We shall now address some of these issues further within the context of experiences acquired at Sunderland.

2.2 The TEED Modelling Project

In 1989 members of staff in the School of Computing and Information Systems were awarded a grant of some £120,000 by TEED, an agency of the Department of Employment of the UK government to investigate the use of a computer based approach to teaching mathematical modelling [1]. The target audience were to be beginning non-mathematicians with a weak mathematics background. The project had two strands. The first involved the use of *HyperCard* and STELLA and the second *HyperCard* and *Mathematica*. In both strands *HyperCard* was used to illustrate modelling scenarios and provide simulations. In the second strand *HyperCard* was also used to undertake the modelling process. Once the mathematical model had been obtained *Mathematica* was used to analyse and solve the model equations, to make predictions about model behaviour and to investigate the effect of varying model parameters. The assumption was that students with poor mathematical skills could use the computer algebra package to carry out the mathematical manipulations and plot graphs of solutions. The following example illustrates the procedure.

One of the models chosen was the motion of a projectile. In this case a shell fired from a gun. The first stage was to present the students with a *HyperCard* simulation in which effects such as muzzle velocity, mass of shell, air resistance and angle of projection could be observed. The next stage examined each of these effects in detail making assumptions about which effects were important in deriving mathematical descriptions. At this stage a mathematical model of the gun-projectile system evolved. *Mathematica* was then used to solve the resulting differential equations. The solutions were then manipulated using *Mathematica* to make predictions about maximum range and optimum angle of projection. This analysis was contained in a *Mathematica* notebook complete with diagrams, textual explanations and *Mathematica* commands. At the end of the notebook the students were presented with a number of different scenarios to investigate. One example was one in which a shot putter puts a shot from shoulder height. The students were encouraged to use *Mathematica* to solve the equations and to make new predictions about maximum range and optimum angle of projection.

Here the computer algebra package was being employed both as a mathematical assistant and as a living textbook. The features of the package being exploited were mainly its symbolic and graphical abilities. Assessment was by open-ended tasks of an exploratory nature.

2.3 The TEMPUS Project

In 1991 the School of Computing and Information Systems was awarded approximately 500,000 ECU over three years under the TEMPUS programme. This was, together with partners in Eire and Germany, to develop mathematical modelling courses for graduate students at the Technical University of Sofia. As part of this programme staff at Sunderland delivered a short modelling course at the Technical University. Part of this course involved the introduction of population models based on the Leslie matrix model. It was decided to use *Maple* to assist in the mathematical analysis and implementation of these models. Using funds provided by the project, a laboratory containing a mixture of Macintosh LCIII's and Quadras had been established at the Technical University. *Maple V* was available on these machines. This version of *Maple* provided an updated version of the linalg package and a version of the programming language with considerably enhanced error diagnostics. After a brief introduction to *Maple* and to population modelling the students used *Maple* to investigate the properties of Leslie models. The Leslie matrix which occurs in these models has interesting mathematical properties which have implications for the behaviour of the model. One of the most important properties is that the matrix has a single positive, real, dominant eigenvalue. In order to prove this it is necessary to obtain a general form for the characteristic polynomial of the matrix. This can be done using *Maple* by obtaining the characteristic polynomial for various orders of matrix using the *charpoly* function. From these results it is easy to deduce the general form for the characteristic polynomial. Another use of *Maple* is to calculate the development of the population over a number of generations. Graphing the results indicates clearly that the growth of individual age groups is exponential. This corresponds to the fact that the matrix has a single real dominant eigenvalue. *Maple* can the be used to deduce the natural growth rate of the population and the limiting population distribution using the *Maple Eigenvect* command which uses the QR algorithm to calculate the numerical values of the eigenvalues and eigenvectors of the Leslie matrix. Here *Maple* is being used as a mathematical assistant. The numerical, symbolic and programming facilities of the package are all being exploited. The assessment of this course involved the students in obtaining a revised Leslie matrix model of the population growth of red deer on the isle of Rhum. This model had to take into account the growth of both the male and female deer populations. This leads to a modified Leslie matrix with properties analogous to those of the original. These properties can be investigated using the linalg package. Once more this assessment is of the extended, unsupervised, exploratory type.

2.4 Computer Algebra Based Courses

The University of Sunderland runs a large combined programme in which students can choose modules from a wide range of disciplines. According to the modules chosen student's can progress to a named degree or take a combined degree in which they can major, minor or take a joint degree in a particular subject area. This programme offers a number of mathematics modules at various levels. In the past mathematics has recruited reasonably well in the first year. However second and third year modules have often had very low numbers of students. One third year module has run with only one student in the past. In addition the best mathematicians were being lost to other disciplines. This position is untenable in the present climate with its emphasis on efficiency and value for money. After considerable discussion it was decided that both the content and style of delivery of the mathematics courses was inappropriate for the type of student recruited onto the degree programme. Mathematics courses, particularly in pure mathematics, had been delivered in a traditional lecture/tutorial mode with considerable emphasis on the theorem/lemma/proof approach. Computer use was minimal. Statistics packages and spreadsheets were used in some of the applied mathematics modules but virtually no use was made of computers in pure mathematics. This style of course delivery was considered inappropriate for students who were not primarily mathematicians and were mainly interested in using mathematics as a tool in their other subjects. As a consequence two new modules, Computer Based Mathematics I and II were introduced to replace pure mathematics and some of the numerical analysis. These modules are totally computer based, using computer algebra packages, with few formal lectures or tutorials. These modules were run for the first time in the 1992-93.

Computer Based Mathematics I used both *Derive* and *Maple*. This decision was made so that at the end of the first year students had experience of both a PC and a Macintosh package. Clearly some students would not take further mathematics modules but would need to use mathematics in other disciplines. Most of the other schools in the university are PC based. As subsequent modules would make use of *Maple* it was felt that this package should be introduced in the first year. Computer Based Mathematics II was completely *Maple* based in order to exploit the more powerful features of that package. The mode of delivery of both these modules was similar. Students were provided with comprehensive handouts which contained an explanation of the mathematical topic which invariably employed *Derive* or *Maple* commands to carry out necessary mathematical operations, exercises for the student to attempt and a final section containing further exercises and more substantial problems of an investigative nature. During class students sat at the keyboard and worked there way through the handout under the supervision of the lecturer. Occasionally the class would be taken in a lecture room if it was considered that a short introduction to a topic was more appropriately delivered by a short formal lecture or towards the end of a topic to discuss any common points of difficulty or interest that may have arisen.

Using computer algebra often implies that a totally different type of exercise must be set. It is no longer sufficient in calculus to present the students with a long list of derivatives or integrals to be calculated. Computer algebra packages make this a trivial matter. Exercises on differential calculus need to concentrate on investigating properties of functions. This can involve looking at intervals where the function is increasing or decreasing, investigating roots and stationary points and investigating continuity and differentiability. The graphing facilities of the computer algebra package can be used to obtain a visual representation of these concepts. This has a number of advantages over a more traditional approach. More realistic problems can be tackled. It is no longer necessary that functions have roots and stationary values satisfying some easily solved equation with integer or rational roots. Many more examples can be tackled as the computer algebra package takes care of time-consuming mathematical manipulations and graphing. The effect of making small changes to the function can easily be illustrated and investigated. Similarly in integral calculus it is possible to concentrate on issues such as integrability and the existence of infinite integrals. Special functions can be introduced at an early stage. In algebra matrices larger than 3x3 are easily handled.

Assessment of the course also has to be handled differently. A possible ideal form of assessment is to set assignments of an investigative nature. However at present course boards and external examiners are reluctant for mathematics courses to contain 100% continuous assessment. Consequently some form of time-constrained test is required. In both Computer Based Mathematics I and II 50% of the marks were awarded for coursework and 50% for a time-constrained examination. In Computer Based Mathematics I the coursework was split between an assignment of an extended, exploratory, unsupervised type together with a shorter time-constrained, supervised task. This latter task was also of an investigative nature. In Computer Based Mathematics II all the coursework was of the extended, exploratory type and in both courses the examination was taken at the keyboard. In Computer Based Mathematics I students could be asked to investigate the continuity or differentiability of a function both analytically and graphically or they might be asked to determine integrability. In Computer Based Mathematics II students could be asked to investigate the nature of the equilibrium points of a system of differential equations and use the *Maple phaseplot* command to justify there findings graphically. This type of time-constrained assessment presents a number of logistic problems. Students are not used to sitting examinations at the keyboard. Time had to be allowed for them to become comfortable with the machine, to make sure that the computer algebra package together with all the relevant packages was loaded and that they could readily save their work. The format in which students would provide their answers had to be considered. Three possibilities were considered. Students could save all their work to disk and the examiner could mark from the file. The problem here is that files do become corrupted. Students could print out their work. Here the problem was time and lack of printing resources. Finally students could fill in a specially prepared answer booklet. One problem in this case concerns the student providing enough information for the examiner to follow through working in the case of an incorrect answer without the student essentially doing the whole problem by hand. Another is in obtaining adequate sketches of graphs. Initially the first option was tried. However this proved to be especially unsatisfactory in the case of *Maple*. The Macintosh version of *Maple* seemed to corrupt files at random, mixing up text, input and output. As yet despite several suggestions and considerable experimentation no cause for this corruption has been discovered. The third option was finally settled upon as presenting the fewest problems.

The results are summarised as follows. The pass rate on Computer Based Mathematics I was similar to that on the replaced course in previous years but on Computer Based Mathematics II rather better. This was confirmed to some extent by informal feedback from the students taking Computer Based Mathematics II. The general opinion of the class was that they preferred the computer based approach to lectures. At first they were apprehensive about sitting a totally computer based examination, but their fears were dissipated on actually taking the examination. Although the new courses could not be claimed to be an unmitigated success there were clear reasons for this. The courses were based on a teaching approach which was untried both by staff and students. Handout material had to be prepared from scratch largely without the aid of textbooks or prior experience. Student exercises and assessments were of a completely different type from those used on a traditional mathematics course and had to be devised largely unaided. Some students undoubtedly experienced computer phobia. Many of these difficulties will lessen as staff become more experienced, as handouts are modified in the light of feedback from students and other institutions teach mathematics using a computer algebra approach leading to the sharing of experiences.

Although it is too early to speculate on whether mathematics can be taught in this way some obvious problems do exist. Students become largely reliant on the package to carry out manipulations and to plot graphs. This has serious implications if later they are expected to do mathematics in an environment where computer algebra packages are unavailable. Computer algebra packages do get wrong results and even experienced mathematicians may not immediately spot the error. For these reasons it is necessary that students still do some calculations by hand and that they are taught to be critical of machine answers. This experience will be built into the course for 1993-94.

3 Dynamical Systems Using Maple

The School of Computing and Information Systems at Sunderland runs two mathematically oriented masters courses, Scientific Computing and Decision Support Systems. These courses take students with a wide variety of prior experience. Students have first degrees in such areas mathematics, engineering, economics and computer science. Both of these courses contain essentially the same dynamical systems module. This module is split into three main areas. The numerical solution of systems of differential equations, phase plane analysis of autonomous systems of differential equations in the plane and an introduction to chaos through non-linear difference equations and non-linear systems of differential equations. *Maple V* Release II running on Sun Sparc stations was used as a numerical, symbolic and graphing tool throughout the module. The use of the package in each of these areas will now be considered.

3.1 The NODE Package

Although *Maple* does include a number of numerical routines related to the numerical solution of systems of non-linear differential equations including a solver based on the Runge-Kutta-Fehlberg method they are inadequate to support the numerical component of the dynamical systems course. For this reason the NODE package was developed at Sunderland. This is a *Maple* external file written in the *Maple* programming language and compiled into *Maple* internal format. The package contains a number of procedures for the numerical analysis and implementation of methods for the numerical solution of systems of non-linear differential equations. For example the implemented methods of solution include second order Runge-Kutta methods, the classical 4th order Runge-Kutta method and the variable step, 4th/5th order, Runge-Kutta-Fehlberg method. Each of these methods is written in such a way as to provide the user with complete control over step-size and error tolerance. This is crucial in, for example, investigating problems of stiffness. One of the routines provided for analysis of numerical methods is the *rootlocus* procedure. This plots stability regions for linear multistep methods using the root locus method. The details of the procedure and an example of its implementation are given below.

The listing below is of the original source file for the procedure. The first section contains the help text which is displayed on invoking the *Maple* help command ? rootlocus. The remainder of the file contains the *Maple* code required to plot the stability region. The parameters r and s are respectively the first and second characteristic polynomials of the linear multistep method.

```
> `help/text/rootlocus` := TEXT(

    `FUNCTION: rootlocus - draw the boundary of

    the region`,
    `of absolute stability for a L.M.M.. with

    first and second`,

    `characteristic polynomials r(z) and

    s(z)respectively`,

    `using the root locus method.`,``,

    `CALLING SEQUENCE:`,`rootlocus(r,s)`,

    `PARAMETERS: r,s - two functions defining

    the first and`,

    `second characteristic polynomials of the

    L.M.M.`,``,``,

    `EXAMPLES`,``,

    `r1:=z->z-1;`,

    `s1:=z>1;`,`rootlocus(r1,s1);`,``,

    `r2:=z->z^2-

    z;`,`s2:=z>3*z/21/2;`,`rootlocus(r2,s2);`);

    rootlocus := proc (r, s)

    local i, w, x, y, arg, pltlst;

    pltlst := [];

    for i from 0 to 50 do

    arg := exp(1/25*(-1)**(1/2)*i*Pi);
```

```
w := evalf(evalc(r(arg)/s(arg)));

x := evalc(Re(w));

y := evalc(Im(w));

pltlst := [op(pltlst), x, y]

od;

plot(pltlst)
```

```
end;
```

Below is an example of an implementation of the method to the 4th order Adams-Bashforth method with characteristic polynomials given by

$$\rho(z) = z^4 - z^3$$

$$\sigma(z) = \frac{55}{24} z^3 - \frac{59}{24} z^2 + \frac{37}{24} z - \frac{9}{24}$$

Firstly we need to load the NODE package. This is done using the *Maple read* command

```
> read `NODE.m`;
```

Next we define the characteristic polynomials using *Maple*'s function notation

```
> r:=z->z^4-z^3;

                4     3
    r:= z -> z  - z

> s:=z->55/24*z^3-59/24*z^2+37/24*z-9/24;

                    3            2
    s:= z -> 55/24 z  - 59/24 z  + 37/24 z - 3/8
```

Now we call the rootlocus procedure

```
> rootlocus(r,s);
```

The resulting stability region is shown in Fig.1.

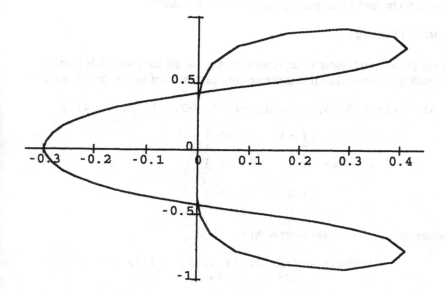

Fig. 1 Stability region of the 4th order Adams-Bashforth method

3.2 Systems of Linear Differential Equations

Before examining non-linear systems of differential equations it is necessary that students have a thorough understanding of the properties of and solution methods for linear systems of differential equations. The *Maple linalg* package can be used to determine the normal modes of such system. Consider the system

$$\mathbf{x}' = \mathbf{A}\,\mathbf{x}$$

We can use the *eigenvect* command to find the complete eigensystem of the matrix **A**. For example consider the system

$$\begin{bmatrix} x' \\ y' \\ z' \end{bmatrix} = \begin{bmatrix} -1 & 1 & -1 \\ 1 & -3 & -1 \\ 2 & 0 & -4 \end{bmatrix} \begin{bmatrix} x \\ y \\ z \end{bmatrix}$$

We need to begin by loading the package using the *with* command

```
> with(linalg):
```

Note the use of the : terminator to suppress a printout of all the commands contained within the package. Now we can find the complete eigensystem of the coefficient matrix

```
> A:= matrix(3, 3, [-1, 1, -1, 1, -3, -1, 2, 0, 4]);
```

$$
A:= \begin{bmatrix}
-1 & 1 & -1 \\
1 & -3 & -1 \\
2 & 0 & -4
\end{bmatrix}
$$

```
>eigensysA: = eigenvects(A);
```

$$
\text{eigensysA}:=[-2, 2, \{[1, 0, 1]\}], \\
[-4,1, \{[0, 1, 1]\}]
$$

From this we can see that the system has two eigenvalues -2 and -4 and hence has stable equilibrium at the origin. The first eigenvalue -2 is an eigenvalue of multiplicity 2 corresponding to the single eigenvector [1, 0, 1]. In this case the exponential matrix method needs to be employed to find the general solution of the system. The *linalg* package contains the *exponential* procedure to find the exponential matrix

$$e^{At}$$

The procedure takes two arguments, the matrix A and the scalar t to represent time.

```
> expAt: = exponential(A, t);
```

$$
\text{expAt}:= [t \exp(-2t) + \exp(-2t), \\
t \exp(-2t), -t \exp(-2t)]
$$

```
    [-1/2 exp(-4t)+ 1/2 exp(-2t),

    1/2 exp(-4t)+ 1/2 exp(-2t),

    1/2 exp(-4t)- 1/2 exp(-2t)]

    [t exp(-2t)- 1/2 exp(-4t)+ 1/2 exp(-2t),

    t exp(-2t)+ 1/2 exp(-4t)- 1/2 exp(-2t),

    -t exp(-2t)+ 1/2 exp(-4t)+ 1/2 exp(-2t)]
```

Now we can construct the general solution of the system using the linearly independent vectors [1, 0, 0], [0, 1, 0] and [0, 0, 1] as a basis.

```
> x: = evalm(C1*expAt&*[1,0,0] + C2*expAt&*[0,1,0] +

    C3*expAt&*[0,0,1]

  x: = [C1 exp(-2t)(t + 1) +  C2 t exp(-2t) -

    C3 t exp(-2t),

    1/2C1(-exp(-4t)+ exp(-2t)

    + 1/2C2(-exp(-4t)+ exp(-2t)

    + 1/2C3(-exp(-4t)- exp(-2t)),

    1/2C1(2t exp(-2t)+ exp(-4t) + exp(-2t))

    + 1/2C2(2t exp(-2t)+ exp(-4t) - exp(-2t))

    - 1/2C3(2t exp(-2t)- exp(-4t) - exp(-2t))]
```

3.3 Phase Space

Maple contains many commands which can aid in the phase plane analysis of systems of autonomous differential equations in the plane. Some of these are generic *Maple* commands for such tasks as solving non-linear algebraic equations and integrating differential equations. Other useful commands are found in the *linalg* package. *Maple* also has a package, *DEtools* which contains commands specifically designed for the phase plane analysis of differential systems. We shall consider a number of possible uses.

Consider the system of differential equations

$$x' = x(1 - y^2)$$

$$y' = y(1 - 2x - x^3).$$

To find the equilibrium points we must solve the simultaneous non-linear equations

$$x(1 - y) = y(1 - 2x - x^3) = 0.$$

This can be accomplished using a combination of the *Maple* *solve* and *fsolve* commands. Firstly we need to define two expressions representing the right hand sides of the equations.

```
> f:=x*(1-y^2):g:=y*(1-2*x-x^3):
```

The command solve finds exact solutions where possible.

```
> eqpts:=solve({f=0,g=0},{x,y});

eqpts:= {x = 0, y = 0},

                            3
{y = 1, x = RootOf(- 1 + 2 _Z + _Z )},

                            3
{x = RootOf(- 1 + 2 _Z + _Z ), y = -1}
```

Note the occurrence of the *Maple* expression *RootOf* here. We can use fsolve to find the real root of this cubic

```
> r:=fsolve(z^3+2*z-1=0,z);

r:= .4533976515.
```

Now we see that there are three equilibrium points (0,0), (r,-1) and (r,1). We can investigate their stability by examining the eigenvalues of the matrix

$$
\begin{bmatrix} \dfrac{\partial f}{\partial x} & \dfrac{\partial f}{\partial y} \\[2mm] \dfrac{\partial g}{\partial x} & \dfrac{\partial g}{\partial y} \end{bmatrix}_{(x_*, y_*)}
$$

where (x_*, y_*) is an equilibrium point. In order to do this we need to load the *linalg* package

> with(linalg):

> A:=matrix(2,2,[diff(f,x),diff(f,y),diff(g,x),diff(g,y)]):

> A1:=subs(x=0,y=0,evalm(A));

$$
A1 := \begin{bmatrix} 1 & 0 \\ 0 & 1 \end{bmatrix}
$$

> evs1:=eigenvals(A1);

    evs1:= 1, 1

> A2:=subs(x=r,y=-1,evalm(A));

$$
A2 := \begin{bmatrix} 0 & .9067953030 \\ 2.616708291 & \dfrac{-10}{.4*10} \end{bmatrix}
$$

> evs2:=eigenvals(A2);

    evs2:= 1.540395660, -1.540395660

> A3   :=subs(x=r,y=1,evalm(A));

$$
A3 := \begin{bmatrix} 0 & -.9067953030 \\ -2.616708291 & \dfrac{-10}{.4*10} \end{bmatrix}
$$

```
> evs3:=eigenvals(A3);
```

```
evs3:= 1.540395660, -1.540395660
```

From these results it is clear that (0,0) is an unstable node and (r,-1), (r,1) are a pair of saddle points. Confirmation of this can be obtained using the command *DEplot2* from the *DEtools* package. This command produces a phase plane plot of the solution trajectories. This is one of the variations of the generic command *DEplot* which is used to plot solution curves for systems of differential equations. Once more we must begin by loading the package

```
> with(DEtools):
```

The command has many variants allowing control over such matters as whether only the direction field is plotted or trajectories are included, the stepsize to be used by the numerical solver and the plot range. An appropriate form in this case which will include a number of trajectories is

```
> DEplot2([f,g],[x,y],0..10,{[0,0.25,1],
          [0,0.25,-1],[0,1,0.25],[0,-1,0.25],
          [0,1,-0.25],[0,1,1],[0,1,-1],
          [0,-1,-1]},x=-2..2,y=-2..2);
```

This produces the plot displayed in Fig. 2. This confirms the nature of the equilibrium points but if further clarification is required the plot range can be altered to home in on a region close to one of the equilibrium points.

Maple can also be used to investigate first integrals. Consider the following system of differential equations [2]

$$x' = 2xy;$$

$$y' = y^2 - x^3$$

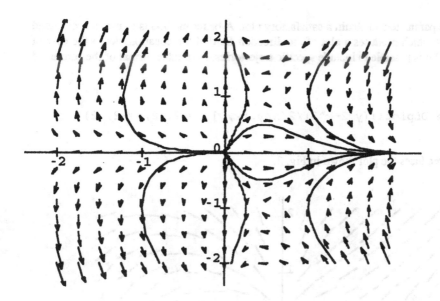

Fig. 2 Using DEplot2 to obtain a phase plot

This system has a non-simple equilibrium point at the origin and hence the linearization theorem is inapplicable. The first integral is obtained by solving the differential equation

$$\frac{dy}{dx} = \frac{y^2 - x^3}{2xy}$$

Although this differential equation can be put into integrating factor form by means of the substitution $u = y^2$ it is easier to let *Maple* do the work. Using the *Maple dsolve* command we obtain

```
> dsolve(diff(y(x),x)=(y(x)^2-x^3)/2/x/y(x),y(x));
```

$$y(x)^2 = -\frac{1}{2} x^3 + x \_C1$$

We could now use the *Maple contourplot* command from the *plots* package to plot the solution curves from the above expression. However it is often difficult to choose suitable parameters to obtain a satisfactory plot. A better method is to use the command *DEplot1* which produces a plot of the direction field of the equation. As in the case of *DEplot2* it is possible to include specific trajectories. A suitable format of the command is

```
> DEplot1((y^2-x^3)/2/x/y,[x,y],x=-2..2,y=-2..2);
```

which produces the plot shown in Fig. 3.

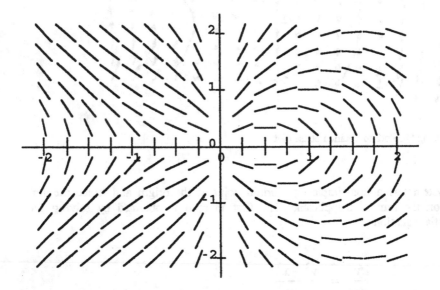

Fig. 3   Plot of the direction field of a system of differential equations

3.4 The Logistic Mapping

The mapping

$$x_{n+1} = a x_n (1 - x_n)$$

arises out of simplified mathematical models of population growth and the economy. The mapping is often used as an introduction to chaos in dynamical systems. Because of the context in which the mapping occurs we usually require that $0 \leq x_n \leq 1$. It is easily shown that this implies that the parameter a must lie in the interval $0 \leq a \leq 4$. The mapping has equilibrium points at $x_n = 0$, $x_n = 1 - 1/a$. It is easy to verify that for the above range of the parameter a the first equilibrium point is stable for $0 < a < 1$ and the second stable for $1 < a < 3$. The most interesting behaviour is when $3 < a \leq 4$. In this interval we can observe a typical period doubling cascade leading to chaos. It is also possible to observe windows where chaos ceases and periodic orbits appear. The *Maple* plot command is very useful in illustrating these effects graphically. An effective method of showing the behaviour of the mapping is the cobweb diagram. This diagram consists of a plot of the straight line $y = x$ and the parabola $y = ax(1- x)$ with superimposed straight lines showing successive iterates. These lines join up the point (x_n, x_n) on the straight line and $(x_n, ax_n(1 - x_n))$ on the parabola. Since we are usually interested in the detection of cycles and chaos it is more convenient to use a simplified version of this diagram in which the straight line is omitted and only the points on the parabola are joined. A sequence of *Maple* commands to accomplish this is given below.

We need the plots package

```
> with(plots):
```

Define the mapping

```
> g:=(x,a)->a*x*(1-x):
```

A procedure to produce the plot

```
> web:=proc(g,x0,a)
        local xold,x1,count:
        dlist:=[ ]:
        xold:=x0+1:
        x1:=x0:
        count:=0:
        while count<400 do
                xold:=x1:
                x1:=evalf(g(x1,a)):
                count:=count+1:
```

```
                    if count>250 then
        dlist:=[op(dlist),op([x1,g(x1,a)])]
            fi:
            od:
            p1:=plot({dlist},style=LINE):
            p2:=plot(g(x,a),x=0..1):
            display({p1,p2});
    end:
```

Now implement the procedure

```
> web(g,0.5,3.56);
```

This results in the plot an eight cycle at a = 3.56 shown in Fig. 4.

A more complete picture of the behaviour of the mapping can be obtained from the bifurcation diagram in which the cyclic values of x are plotted against the parameter a. The following *Maple* procedure can be used to obtain this plot

```
> bifurcation:=proc(g,x0,amin,amax)
        local xold,x1,count,tdlist,dlist,range;
        dlist:=[ ];
        a:=amin;
        range:=amax-amin;
        while a<amax do
            xold:=x0+1;
            x1:=x0;
            count:=0;
```

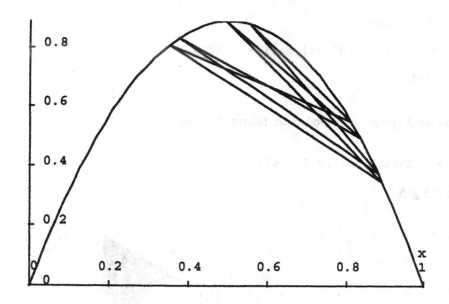

Fig. 4 An eight cycle of the logistic mapping

```
tdlist:=[ ];
      while count<400 do
            xold:=x1;
            x1:=evalf(g(x1,a));
            count:=count+1;
            if count>250 then
  tdlist:=[op(tdlist),op([a,x1])]

            fi;
            od;
            dlist:=[op(dlist),op(tdlist)];
            a:=evalf(a+range/100);
```

```
        od;

        plot({dlist},style=POINT);

    end:
```

We now use the procedure to produce the bifurcation diagram

```
    > bifurcation(g,0.2,1,4);
```

shown in Fig. 5.

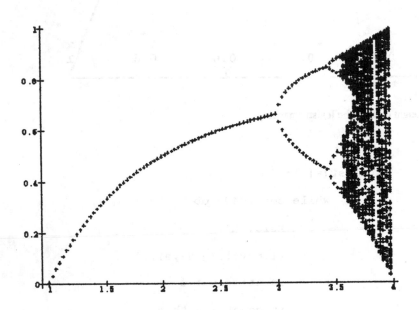

Fig. 5 Bifurcation diagram for the logistic mapping

3.5 Non-linear Differential Equations and Chaos

The Rössler equations [3]

$$x' = -y - z$$
$$y' = x + ay$$
$$z' = b + z(x - c)$$

are often used as a prototype for illustrating chaotic behaviour. The system represents the minimum necessary for chaos to occur in that there are three differential equations and they are non-linear because of the presence of the xz term in the third equation. The behaviour of the system is determined by the parameters a, b and c. The parameter c plays the major role in the generation of chaotic solutions.

The equilibrium points of the system are easily shown to be

$$z = \frac{c \pm \sqrt{c^2 - 4ab}}{2a}$$

$$x = az, \qquad y = -z$$

and clearly for an equilibrium point to exist we require that $c^2 \geq 4ab$. The coefficient matrix of the linearised system is

$$\begin{bmatrix} 0 & -1 & -1 \\ 1 & a & 0 \\ z_e & 0 & az_e - c \end{bmatrix}$$

where z_e is the z-co-ordinate of the equilibrium point. Since the investigation of the behaviour of the system for all possible parameter values is involved and c is the key parameter we shall assume that $a = b = 1/5$. Clearly then for equilibrium points to exist $|c| \geq 2/5$ and since c is normally assumed to be positive we investigate the behaviour of the system for $c \geq 2/5$. Now the eigenvalues of the matrix are complicated functions of c. In order to analyse their behaviour we can use Maple to obtain symbolic expressions for them. The *Maple* plot command can then be used to plot the real part of each eigenvalue for $2/5 \leq c \leq 10$. From these plots it is clear that for both equilibrium points the linearization has one real and two complex eigenvalues. In the case of the equilibrium point where the + sign is taken the real eigenvalue is positive and tends to 0.2 as c increases. The complex eigenvalues have small real parts which are negative but tend to zero as c increases. The other eigenvalue has a real root which is negative and tends to minus infinity as c increases. The complex eigenvalues have a positive real part which tends to 0.1 as c increases. From this we see that both equilibrium points are unstable. We would expect, from the nature of the eigenvalues, that for large values of c some sort of cyclic behaviour would occur near to the first equilibrium point and an unstable spiral near the second. It is the interplay between these two equilibrium points which determines the nature of the trajectories. The *Maple DEplot* command can be used to obtain plots of the trajectories in three space dimensions. An appropriate sequence of *Maple* commands to obtain such a plot is shown below.

```
> with(DEtools):

> a:=0.2:b:=0.2:c:=8:

  f:=-y-z:

  g:=x+a*y:

  h:=b+z*(x-c):

> DEplot([f,g,h],[x,y,z],80..120,{[0,1,1,1]},

  stepsize=0.01);
```

Notice that the trajectory is plotted over the range $80 \leq t \leq 120$. This ensures that the transients have had time to die out. The resulting plot is shown in Fig. 6.

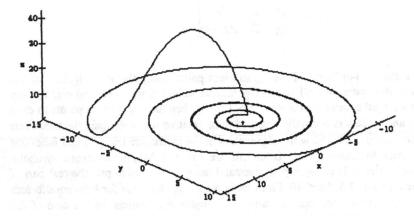

Fig. 6 Phase trajectory of the Rössler system

Another set of differential equations which is much studied in the context of chaos is the Lorentz system [3]

$$x' = \sigma(y-x)$$
$$y' = r x - y - x$$
$$z' = x y - b z.$$

In this system the important parameter is r. The other parameters are usually assigned the values $\sigma = 10$, $b = 8/3$.

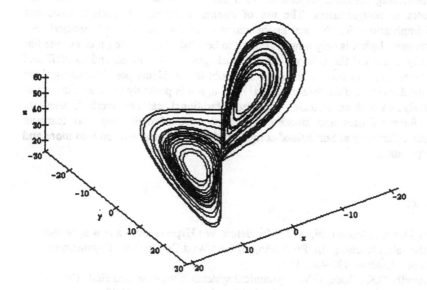

Fig. 7 The Lorentz attractor

The system of equations has three equilibrium points

$$(0, 0, 0)$$

$$\left(\pm\sqrt{b\,(r-1)},\ \pm\sqrt{b\,(r-1)},\ r-1 \right) r > 1,$$

where the origin is easily shown to be unstable. The other two equilibrium points can be shown to be stable provided that $r < r_c$. It is possible to carry out a numerical investigation of the eigenvalues as was done for the Rössler system. This reveals that for these parameter values $r_c = 24.74$. For values of $r_c > r$ the system usually exhibits chaotic behaviour with solution trajectories lying on a strange attractor. A plot of the attractor can be obtained using *DEplot*. An example of such a plot is shown in Fig. 7.

4 Conclusions

This paper has outlined a number of ways in which a computer algebra package can be used in delivering mathematics courses to a range of students from first year undergraduates to postgraduates. The use of computer algebra in such courses has profound implications for the way mathematics is taught, learned and assessed. Because computer algebra is only now beginning to be used as a teaching tool no one has satisfactorily addressed these issues. As the packages are developed and as staff and students become experienced in their use no doubt the problems posed by using them will be solved and new directions explored. Although it is probably the case that use of computer algebra will never totally supersede traditional teaching methods they are going to force teachers and students into new ways of delivering and learning mathematics. Certainly at Sunderland computer algebra is being written into more and more programmes.

References

1 Curran, D.A.S., Moscardini, A.O., Middleton, W.: Hypermedia - a new approach to mathematics teaching. In: Proceeding of East West Congress on Engineering Education, Cracow, 432-436, 1991
2 Arrowsmith, D.K., Place, C.M.: Dynamical systems, Chapman and Hall, 1992
3 Drazin, P.G.: Non-linear systems, Cambridge University Press, 1992.

Design of Mathematical Modelling Courses for Fluid Flow in Engineering Education

C. Çiray

DAE, Middle East Technical University, 06531 Ankara, Turkey

Abstract. The main body of thoughts forming the backbone of the exposition "Design of Mathematical Modelling Courses for Fluid Flow in Engineering Education" stem from and directed to two kinds of people who can benefit from such an exposition. These two groups are the users of ready made codes or computer programmes and those who compose or prepare such codes. Following is an introduction explaining in essence the exposition and various concepts and activities related to mathematical modelling. The second section is the main section divided in subsections and attempts to lay down the sequences and the logic in mathematical modelling of fluid flow. The hierarchical approach may be visualised as an attempt to classify mathematical modelling and the its teaching, though it appears to be impractical to give a classification (a kind of branch tree). The second section discusses at the same time what may be thought at various levels of engineering education in terms of mathematical modelling of fluid flow, rather briefly at graduate level and above in broad terms, but more in detail at undergraduate level. The exposition ends with a brief conclusion.

Keywords. Mathematical modelling, engineering education, fluid flow, computer, paradigm

1 Introduction

The background and scope of the NATO-ARW on Mathematical Modelling (MM) underlines and calls for attention the need of proper teaching of MM in the basic engineering sciences, namely Fluid Mechanics (FM) and Heat Transfer.

Generally accepted mathematical continuum model of fluid flow is essentially complex even for simple problems of real nature. Consequently the need to use computers in the computational efforts to obtain results for engineering applications became an inseparable activity of the engineers. The activities in such an engineering process, as related to the use of computers, involves two fundamental categories: first, the realisation or composition or writing the software, and second, the use or application or running an existing software for a given engineering problem. In addition, the advances in computer hardware do not only lay the way for more realistic, accurate, optimised, economical and quicker engineering solutions, but they also lay down the requirements and expectations for such quality software. Therefore within this Workshop on MM Courses for Engineering Education it is intended rightly to address to the specific area of fluid flow, to the teaching of FM since the latter can be interpreted as teaching the MM of fluid flow to engineers who are likely to be the users or composers of the software, hence MMs.

A presentation on the organisation of a course on MM can be prepared at least in two ways. In the first one, one can convey his ideas by giving an example of his choice, *i.e.* while explaining the construction of a mathematical model of a given natural phenomenon, the person can also generalise his ideas in relation to the process of organising a course on MM. The second approach can concentrate on the analysis of MM, to the requirements from a fundamental point of view and in a general manner right from the beginning. The presentation in hand adopts the second approach, though, looking at the titles and abstracts available at the moment of preparation of these ideas, the second approach is common in the workshop papers.

MM is one of the steps in a series of activities directed to the use of mathematics for the purpose of scientific studies or engineering applications. The ultimate mathematical result of the problem may come out in terms of implicit or explicit algebraic formula (or formulae) or the result may have to come out in numerical form through an analog or digital process using appropriate hardware. The steps involved may be enumerated as follows:

> *(a): the definition of the actual problem*
> *(b): the choice of the paradigm (physical model, idealised)*
> *(c): the mathematical modelling*
> *(d): the choice of solution methodology*
> *(e): the solution, numerical results*
> *(f): interpretation.*

In spite of the fact that the scope of this workshop is limited to MM education for engineers, [item (c)], which is obviously the subject of this paper, items (a) and (b) will be touched to certain degree in the second section of this paper. Therefore it is felt appropriate to say few words about items (d), (e) and (f) before embarking on the actual topic.

The choice of the solution methodology can lead either to the so-called analytic solutions, meant to be closed form or algebraic formulae, or one may be led to use analog or digital computers. Because of the nature of fluid flow problem, with its reflection on the paradigm and consequently on the mathematical model, analytic solutions are either very (very) scarce, or non-existent or extremely restricted in application scope for realistic problems of interest to engineers.

The use of analog computers is restricted in the scope of its use and versatility compared to digital processors, but may be efficient and cost effective for a specific problem or a class of problems.

In the case of use of digital computers, the whole process is named computational fluid dynamics (CFD) if the problem is a fluid flow problem. In such an approach, the solution methodology involves:

- *discretization*
- *construction of the algorithm*
- *programming*
- *debugging, and last but not the least*
- *validation.*

Many of these have their counterparts for analog computers and even for analytic solutions where possible. Any of these steps in the solution process influences the decision on the of what the next step has to be or can be. For example if the actual problem in hand is a laminar one, a paradigm and a mathematical model of a turbulent flow may not be used (perhaps should not be used). Or, a mathematical model and the associated programme should not be expected to give viscous drag if a panel method is used (validation).

Once the solutions are completed and numerical results are obtained [step(e)], the interpretation [step (f)] follows. With vast number of numerical results coming out from the computers, the graphical representation or the post processing becomes critical and turns out to be an inseparable tool of the interpretation.

The presentation in hand will concentrate on the step *"MM: mathematical modelling': i.e.* item (c), but the question of paradigm [item (b)] is tightly related to the MM, therefore it will be given some consideration in the sequel. Yet it is to be noted that all of these steps are also important to the engineer and adequate weight must be allocated to each of them during a proper CFD course. The need for attention to these facts are warranted. Indeed, the rapid increase of hardware capabilities in terms of memory, and speed of calculation is breathtaking. Emmons [1] estimated that the direct simulation of a turbulent pipe flow at a Reynolds number of 5000 requires approximately 10^{14} operations. The speed of an operation was given as 10 μs, or 0.1 megaflops for the computers at that time of the publication of the paper, in 1970. Thus the time needed for the direct simulation of the turbulent pipe flow at Re = 5000 was estimated to be about 100 (a hundred) years. Today, people talks of computer speed of giga flops which means an increase of 10000 folds. The same problem, therefore, may be expected to be solved within few days. In terms of memory capacity, a million of grid points each having six unknowns are problems that are handled in aeronautical industry. Such vast capabilities incur more responsibilities to engineers in order to make a sensible use of the capabilities with which they play.

It is the belief of the author that the engineers, whether they are users or composers, they must be thoroughly educated to the fundamentals of the steps (a) to (f) mentioned above in order to compose software commensurate and adequate for the class of fluid flow problems of interest and to be aware of the limitations of the capabilities of the software in their hands.

2 Mathematical Modelling and Fluid Flow

2.1 Modelling the Flow

The mathematical model is essentially a mathematical expression or an algorithm which when solved or performed gives results that can be used to find quantities to predict or to assess the behaviour of a fictitious fluid flow phenomenon that is claimed to be equivalent to the fluid flow problem as it occurs in nature.

So, first, there is the actual fluid flow problem as it occurs in nature, secondly the paradigm or the idealised phenomenon that must be equivalent to the natural one conceived by the composer and third the mathematical expression that relates various quantities involved in the paradigm.

The term mathematical expression is meant to be understood as the collection of all kinds and all of the mathematical expressions, *i.e.* algebraic, differential, integral *etc.*, and all of the boundary and/or initial conditions that are needed to define mathematically the relations between the variables involved in the idealised phenomenon or the paradigm. Some time a series of steps or a *"scenario "* is devised as the paradigm to take the place of the actual phenomenon in nature. Then, an algorithm containing mathematical expressions organised in conformity with the scenario is used, and this becomes the mathematical model.

Examples of the latter can be reminded from Rarefied Gas Dynamics (RGD) that are grouped under Molecular Dynamics (MD) and Direct Simulation Monte Carlo (DSMC) methods [2], [3].

It may be proper to underline at this stage, that any available computer code solves the idealised fluid flow problem rather than the actual fluid flow problem that the code user has in mind.

2.2 Paradigm Level

The paradigm is an idealised-simplified counterpart of the real phenomenon in nature conceived to an extent of complexity that can reflect the characteristics at an accepted level (accuracy) that the composer has in mind.

If the aim is to predict the pressure distribution, therefore the lift and the pitching moment for an airfoil subject to an unseparated flow of an incompressible fluid within an error margin not exceeding 10%, the paradigm used is a continuum representation of the actual fluid with a fluid of constant mass density, without any viscosity and with a flow which is planar and irrotational from kinematic point of view. Such a paradigm will give rise to a mathematical model that can predict pressure distribution and the associated integral quantities to an acceptable accuracy. But the user should not expect to obtain the stall angle, the wall shear stress distribution or the drag from the output of this programme. If pushed to calculate the drag, it is likely to find some numbers, but it is clear that these numbers can not have any relation to the drag that the user has in mind.

The paradigm related to fluid flows are in two categories. The first comprises paradigms representing the fluid at molecular level with an aim to assess the behaviour at macroscopic level. The second category considers the fluid as a continuum and ignores the molecular structure of the fluid.

The representation of the fluid at molecular (or even atomic) level considers the motions that are associated with the molecules. The general name associated with this kind of representation is cellular-automata models and lattice-gas cellular-automata models are reviewed in [3]. Various paradigms differ in lattice geometry and the restrictions imposed on the molecules (particles) in terms of their motions. These paradigms are far from replicating the molecular activity of the real gas. But brought down to suitable small cell sizes, the gross characteristics calculated by averaging procedures replicate the real gas behaviour at acceptable level. Discrete paradigms at molecular level are practically used for relatively low Knudsen numbers, *i.e.* Mrefied conditions. It seems that these paradigms can also be used for high Knudsen numbers. Indeed Frisch *et al* [4] explains a paradigm for 2D flows whose results are compared with Navier-Stokes solutions. Discrete paradigms are normally conceived for gases and the author did not come across with one for liquids.

The second category of paradigms considers the fluid as a continuum and ignores the molecular structure and the molecular activity of the fluid. In this case the elementary volume, hence the flow domain is filled continuously with a fictitious substance. The centroid of the elementary volume is associated with fluid and flow properties which reflect either average properties of the molecules that at the moment under consideration are within the elementary volume or properties that represent the consequences of the molecular activity within the same volume. An average property may be the mass density or the velocity vector, whereas pressure, temperature or viscosity can be mentioned as examples of the latter group.

The continuum representation of the fluid flow is used in an hierarchy of complexities reflecting and responding to the complexity of the actual physical problem in hand. The complexity becomes apparent with increasing number of fluid-flow properties associated with the elementary volume as dependent variables and are depicted below:

(1): pressure
(2): pressure and velocity
(3): pressure, density(ies) and temperature
(4): pressure, density(ies), temperature and velocity
(5): previous plus electromagnetic quantities
(6): previous with the addition of chemical reactions leading to consider specific heat ratio as dependent variable.

The first line depicts the simplest type of fluid problem, the no-velocity flow of an incompressible fluid, i.e., the hydrostatic problem. It is understood that the density is given. The complexity brought in by the dependent variables for problems depicted in (5) and (6) lead to a variety of flows studied within magnetofluid dynamics, plasmas, high enthalpy flows *etc.*.

The second item which creates complexities in the continuum paradigms is the fluid itself. The nature of the fluid considered in the paradigm influences the mathematical model. This question will be taken in MM below. Examples like, inviscid fluid, Newtonian fluid with constant or temperature dependent viscosity etc., may help to make clear the idea. This group can be put under the heading of constitutive equations or laws of the fluids considered.

A third item that characterises the continuum paradigm is whether it is representing an elementary volume or an integral volume and further if it is viewed as a system or control volume.

Last but not the least an item bearing on the complexity of the paradigms the regime of flow, *i.e.* whether the flow is laminar or turbulent or in between. This item influences the nature and the number of the dependent parameters and require additional paradigms (hence mathematical models) for turbulence quantities.

No doubt, steadiness or unsteadiness and the dimensionality of the flow should be remembered to complete the items leading to various complexities.

Hence, various groups that must be considered in paradigms are:

    (a): time dependency
    (b): dimensionality
    (c): regime of the flow
    (d): constitutive laws adopted
    (e): fluid-flow variables (items 1 to 6 above)
    (f): size and type of the paradigm (integral, control volume)
    (g): relativistic or non-relativistic motion

2.3 Mathematical Model

The paradigm, as explained above, is the idealised physical model that is supposed to be able to help to analyse or predict the characteristics of the actual problem in nature that the engineer is interested. The mathematical model is expected to give the mathematical relationship between the variables specified in the paradigm. Whether it is a molecular or a continuum paradigm, the mathematical relations between the variables are based on some laws of nature. Usually expressed as conservation principles, they increase in number as the number of fluid flow dependant variables increases. For the category of problems involving only pressure or pressure and velocity [flows (1) and (2)], only mechanical principles, *i.e.* conservation of mass and momentum are used. With the addition of the temperature[flows (3) and (4)] conservation of energy or first law of thermodynamics is added. For problems involving electromagnetic effects, Faraday laws and conservation of charge (continuity equation have to be considered [flows (5)] Finally for chemically reacting flows [flows of group (6)] principles of chemical reaction kinetics must be added.

These principles give rise to equations which relate the dependent variables to some extent, but generally they are not enough to solve the problem in hand since the number of equations and the number of dependent variables do not match. Therefore additional relations of group (d) above must be added to equations coming from the conservation laws in order to equalise the number of dependent variables and equations.

The fundamental and general character of the conservation principles *(basic laws of nature)* and the special and the restricted character of these additional relations (sometime named *subsidiary laws)* must be noted. This is one of the critical issues which influences perhaps the most the validity of the mathematical model checked against the experiment when the experiment is devised to satisfy the gross features of the model such as time dependence, dimensionality, boundary conditions (geometry and the conditions at the boundaries), *etc.*

The combination of subsidiary and basic laws give rise to governing equations such as Navier-Stokes equations, Maxwell equations or stochiometric equations in case of continuum paradigms and to Boltzmann equations in case of molecular paradigms.

The exposition will continue on an example basis for fluid flows of groups (I) to (4) *i.e.* those fluid flows (single specie) excluding electromagnetic effects and chemical reactions in non-relativistic conditions. The paradigm used in this case is a continuum at infinitesimal scale of a thermomechanical problem. Therefore

- *conservation of mass*
- *conservation of momentum and*
- *first law of thermodynamics*

are the basic laws used to start the construction of the mathematical model. The above principles, in case of a control volume approach, lead to:

$$\rho_{,t} + (\rho u_i)_{,i} = 0 \tag{1}$$

$$\rho(Du_i / D_t) = f_{Bi} + \sigma_{ij,j} \tag{2}$$

$$\rho D(e + u_i^2 / 2) / Dt = \rho\dot{q} - (pu_i)_{,i} + \rho(f_{Bi}u_i) + \dot{Q}_{cond} + \dot{W}_{shear} \tag{3}$$

These equations are quite familiar to the reader. The first is the differential equation of continuity, the second one is the Cauchy stress equation written in indicial form for the i-th component and the last one is the differential form of energy equation. ρ, u_i, f_{Bi}, σ_{ij}, e, $\dot{q}$, p, $\dot{Q}_{cond}$, $\dot{W}_{shear}$,,are respectively the density, i-th component of velocity vector, i-th component of body force per unit volume, ij component of the stress tensor, internal energy per unit mass, rate at which heat is added by conduction at the boundaries per unit volume and finally rate at which work is done by shear stresses, again, per unit volume. Summation convention is to be used on repeating indices, a comma represents partial differentiation either with respect to time (t) or to the i-th direction of the coordinate system. D../Dt shows total or substantial derivative.

The equations above can hardly be used to obtain the mathematical relations of a fluid flow problem directly, but govern the flow problem of any fluid as limited by the paradigm leading to these equations. Therefore they can be used for the construction of mathematical models for a variety of fluid flow problems. Therefore, in a sense, they are quite general. Yet to arrive at these equations, a number of conditions are set; therefore in another sense they are quite restricted. Indeed to obtain them:

- *non-relativistic velocities*
- *continuum flow*
- *single specie fluid*
- *thermomechanical limitations*
- *control volume approach*
- *infinitesimal elementary volume*

conditions are a priori accepted.

In order to arrive to a mathematical model suitable to give direct relations between the variables of the fluid flow, subsidiary relations are needed. Constitutive equations relating the stress tensor to flow and fluid properties define classes of fluid flow and when combined with Cauchy stress equation help to come closer to the mathematical model of that class of fluid flow. As an example the assumption of Newtonian fluid will be made in order to continue with the exposition. In this case the constitutive equation is given as:

$$\sigma_{ij} = -(p + 2/3\,\mu\theta)\delta_{ij} + 2\mu e_{ij} \tag{4}$$

where θ, e_{ij}, μ and δ_{ij} are respectively the dilatation (trace of the velocity gradient tensor), rate of strain tensor, the coefficient of dynamic viscosity and the Kroenecker delta. This relation is specific to a class of fluids which form a subclass of another class of fluids. Namely the Newtonian fluid given above is a linear Stokesian fluid with the additional Stokes assumption that the mean normal stress is the thermodynamic pressure assumed to get rid of one of the two Lame constants. With the above relation, the Cauchy stress equation turns out to be :

$$\rho\, Du_i/Dt = fBi - (p - 1/3\ \mu\theta),_i + \mu u_{i,jj} \tag{5}$$

the well known Navier-Stokes equation. The equation of continuity, the Navier-Stokes equation and then energy equation have to yield the mathematical relations between the variables of the thermomechanical flow problem of a Newtonian fluid. It is necessary to use additional subsidiary relations to specify e in terms of temperature T, hence the equation of state and finally heat conduction relation, *i.e.*:

$$e = c_v \, T; \qquad\qquad p = \rho \, R \, T; \quad q = k \, T_{,\,i} \;. \qquad\qquad (6\text{-}8)$$

It is clear that all these subsidiary relations are specific to a class fluids and sometimes are called Navier-Stokes fluids.

Now the mathematical model for the Navier-Stokes fluid is complete if the continuity, Navier-Stokes and energy equations are complemented with the boundary / initial conditions and relations for e, p and q.

More specialized mathematical models can be generated from the above by restricting the character of the fluid (incompressible, in viscid, or both, *etc.*), or the flow (planar, irrotational, uniform, boundary layer, *etc.*). The regime of the flow may be laminar or turbulent. If laminar, the above equations model faithfully the fluid behaviour. If turbulent, additional turbulence modelling is necessary to circumvent the closure problem. The turbulence modelling needed for a specific class of turbulent flows is as critical as the mathematical model of the mean flow equations *i.e.* the RANS equations.

The boundary and/or initial conditions are considered as a part of the mathematical model, since these conditions are needed to define mathematically in a complete manner the relations between the dependent variables of the fluid flow problem.

It may be summarised that, once the paradigm is specified, the basic laws and subsidiary laws (perhaps including turbulence modelling) are used to build the mathematical model. The restrictions on the paradigm, basic laws and subsidiary laws make the model rather specialised to a class of fluid flows.

3 Mathematical Modelling and Engineering Education

Engineering education concept of today calls for a more through teaching for engineering sciences: engineering mechanics, thermodynamics, electrical sciences, material sciences, transport phenomena (fluid mechanics). A typical example is the Core Curriculum given in Appendix A. Any fluid mechanics course can be considered a mathematical modelling course of fluid flow. With such a view, all the aspects and implications of devising paradigms and the ensuing MM must somehow be conveyed to the potential engineer. Three questions arise to reach this aim:

- *what must be given at undergraduate level?*
- *what must be given at graduate level?*
- *what the engineer has to acquire by himself?*

The answer to the third one is easier than the first two: what cannot be given at educational institutions must be left to the engineer.

The second question is a wide one and depends on:
- *the level of development of fluid mechanics in the world;*
- *the level of development and the needs of the industry or of the country;*

- *the capabilities of the institution of education in terms of facilities and scientists;*
- *the interest of the educating people and those to be educated.*

The graduate level education is tightly connected to the research activities. Therefore scientific and engineering research requirements of the industry, and/or of the country are the deciding factors. If the country is not interested in hypersonic vehicles, or high enthalpy flows, research facilities either do not exist or are not operational in that country. It may be impossible to find support (financial) in this area; hence graduate area in the field will not go beyond individual interest, therefore restricted in scope. On the other hand, if an institution does not have the means to accede to powerful computers, education on direct simulation of turbulence will not be effective.

Graduate courses related to fluid flows normally contain MM or models of some particular flow(s) in few sections of the course. The remaining of the course is devoted to grids, discretization, solution techniques, validation and some examples. These courses appear more in an *"applied nature"*, *i.e.* for specific flows and it is inferred that they are needed and they continue to be fashionable for some time to come. They will be replaced when models of greater generality will be available.

It may be advisable to give a course specifically on MM, perhaps to convey:

- *various paradigms*
- *related basic laws*
- *related subsidiary laws*
- *some examples.*

The above approach may be used at undergraduate level for continuum paradigm within FM course. An example is appended. In the 1st Chapter the continuum paradigm is explained. The 2nd Chapter contains kinematical aspects. Therefore the paradigm is completely pictured in the mind of the student in the first two chapters. The 3rd Chapter is devoted to some fundamentals related to basic laws of nature. In the 4th Chapter the basic laws of nature are transformed into equations that may be used in Mathematical Modelling This course is thought to Mechanical, Civil and Aeronautical Engineering students for long years with slight modifications. The version in the Appendix treats the fluid statics in the 5th Chapter where the constitutive equation is the simplest. Chapter 6 is on Newtonian fluids, and rather simple forms of laminar and turbulent flows are given.

This approach seems to give to the student an awareness of the limitations of specific mathematical models that are used frequently and at the same time a direction for creativity.

4 Conclusion

The capabilities due to computers make it necessary to educate the engineering student more and more to the fundamentals of mathematics and engineering sciences therefore of fluid mechanics. MM of fluid flow is a part of fluid mechanic courses. The education on MM of fluid flow must be organised with an aim to create an awareness of the limitations and also to lead the way for creativity.

Acknowledgement. The author appreciates discussions with Prof. Y. Gogus from Aeronautical Engineering Department, and Prof. C. Toker from Electrical and Electronics Engineering Department of METU. The author would like to thank Mr. U. Arkun, Mr. B. Alkislar and Mr. L. Gokkus for their help in preparing the view graphs and producing copies of the manuscript.

References

1. Emmons, H.: Critique of Numerical Modelling of Fluid Mechanics Phenomena. Annual Review of Fluid Mechanics 2, 15-36 (1970)
2. Bird, G. A.: Perception of numerical methods in rarefied gas dynamics. In: E. P. Munt *et al* (eds) Rarefied Gas Dynamics: Theoretical & Computational Techniques; AAIAA Progress in Astronautics and Aeronautics 118, 211-226 (1989)
3. Boris, J. P.: New Direction in CFD. A.R.F.M. 21, 345-385 (1989)
4. Frisch, U. *et al*: Lattice gas automata for Navier-Stokes equation. Phys Review Letters 56, 1505-8 (1987).

APPENDIX A: METU-FACULTY OF ENGINEERING
                    UNDERGRADUATE CORE CURRICULUM

A. Mathematics and Basic Sciences (min: 9 courses/ 31 credits)

Al: Mathematics (min: four courses/ 14 credits)
* *Compulsory Courses*
• Calculus I;   • Calculus II;   • Differential Equation

* *Elective Courses* (one course in the following areas)
[ ] Linear Algebra; [ ] Complex Variables; [ ] Introduction to Numerical Methods;
[ ] StatisticsandProbability; [ ] Advanced Calculus; [ ] Mathematical Analysis

 A2: Basic Sciences (min: 3 courses / 14 credits)
* *Compulsory Courses*
• General Physics I;   • General Physics II;   • General Chemistry

* *Elective Courses* (one course in the following areas)
[ ] General Physics; [ ] Optics; [ ] Classical Mechanics; [ ] Modern Physics;
[ ] QuantumPhysics; [ ] Astronomy; [ ] General Chemistry; [ ] Analitical, Physical,
Organic Chemistry; [ ] Earth Sciences; [ ] Environmental Sciences; [ ] General Biology;
[ ] Microbiology

B. Engineering Sciences (min: four courses / 12 credits)
(The courses are in 3 of the following 6 areas with at least one course outside the discipline)
[ ] Engineering Mechanics; [ ] Thermodynamics; [ ] Electrical Sciences;
[ ] Material Sciences; [ ] Transport Phenomena; [ ] Computer Sciences

C: Basic Engineering (min: two courses/ 6 credits)
C1: Engineering Graphics (min: one course/ 3 credits)
C2: Computer Literacy & Programming (min: one course/ 3 credits)

D: Communication, Economics, Social Sciences, Humanities (min: four courses/
      12 credits)
Dl: Communication (min: 2 courses/ 6 credits)
* *Compulsory Courses*
• Reading and Writing in English I; • Reading and Writing in English II;
• Advanced Reading in English; • Advanced Writing in English
 D2: Economics, Social Sciences and Humanities (min: two courses/ 6 credits)
(Two courses in the following areas)
[ ] Linguistics; [ ] Foreign Language Studies; [ ] History; [ ] Psychology; [ ] Sociology;
[ ] Philosophy; [ ] Literature; [ ] Fine Arts; [ ] Political Science; [ ] Economics

E: Electives and Design (min: six courses/ 18 credits)
El: Technical Electives (min: four courses/ 12 credits)
E2: Free Electives (min: one course/ 3 credits)
E3: Design (min one course/ 3 credits)

AEE 241: FLUID MECHANICS

1. Introduction
(1) Fluid:definition, description, comparison with solids (2) Fluid mechanics: definition, scope (3) Physical properties: density, viscosity, compressibility (4) Continuum: concept, Knudsen number

2. Kinematics of fluid motion
(1) What is kinematics? (2) Lagrangian and Eulerian coordinates (3) Derivatives: local, convective, total (4) Rates of deformation: velocity, acceleration, rate of volumetric deformation, rate of rigid body rotation, rate of angular deformation, rate of strain (5) Circulation (6) Flow lines:pathline, streamline, streakline

3. Basic principles and methods of analysis
(1) Laws of nature: basic and subsidiary laws (2) Properties:extensive and intensive properties (3) System and control volume approaches (4) Reynolds transport theorem

4. Governing equations and their applications
A. Conservation of mass
(1) Equation of continuity for integral C.V. (2) Equation of continuity for infinitesimal C.V. (3) Planar and axisymmetric flows: stream function
B. Conservation of linear momentum
(1) Force and distributed forces: line forces, body forces,surface forces, stresses, pressure (2) Conservation of linear momentum: for integral C.V., for infinitesimal C.V., Cauchy stress equation
C. Conservation of energy
(1) First law of thermodynamics (2) Energy equation for integral C.V. (3) Bernoulli equation

5. Statics of fluids
(1) Equation of fluid statics: state of stress in fluid statics; equation of fluid statics; pressure distribution in incompressible fluids (2)Equipotential surfaces: manometry (3) Standard atmosphere: lapse rate; temperature, pressure, density distribution in ISA conditions (4) Stability of atmosphere: inversion

6. Flow of real fluids
A. Newtonian fluids
(1) Constitutive equation (2) Navier-Stokes equation (3) Mathematical definition of incompressible fluid flow: equation of continuity; Navier-Stokes equation; boundary conditions (4) Extension to compressible fluid flows (5) Bernoulli equation as the line integral of N.S. equation (6) Internal and external flows
B. Laminarflows
(1) Poiseuille flow (2) Frictional losses
C. Turbulentflows
(1) Turbulent flow characteristics (2) Turbulent flows in straight circular pipes (3) Velocity distribution: law of the wall (4) Head loss calculations: Darcy-Weissbach formula; friction factor, Moody chart (5) Local losses (6) Network principles.

PART TWO

Engineering Mathematics Education

The emphasis now switches to the more general area of engineering education. The scene is set by the paper by Professor Ersoy which identifies mathematical modelling and new teaching methods as two important areas for study. Both areas rely heavily on the use of computers and suitable software.

Mathematical modelling courses were covered in Part One. This part concentrates on software for teaching and as a modelling tool. The papers by Professor Simons and Dr. Pohjolainen cover the use of computer algebra packages, in particular *Mathematica* and *Matlab* respectively. The advantages of courses built around such software is that they very soon become student centred and allow for creative thinking. They thus make an excellent complement to a mathematical modelling course.

The papers by Professors Cecchi and Mikhailov present innovative approaches to the teaching of Finite Element Analysis. Both these authors are very experienced in teaching Finite Elements and the papers emphasise the universality of the method. They present two different global approaches which are then applied to a variety of examples. In contrast to this approach, there is a paper by Dr. Bush on Computational Fluid Dynamics using finite difference methods.

This part finishes with two short papers by Professors Bausch and Flanders. They discuss other computer software that would be useful for engineers.

Integrating New Information Technologies into Engineering Mathematics Curricula

Y. Ersoy[1] and A. O. Moscardini[2]

[1]Middle East Technical University, Ankara, Turkey
[2]University of Sunderland, UK

Abstract. Besides the two major chances world-wide in mathematics education at all levels of schools in almost all developed and developing countries for the last three decades, two recent developments have contributed to the debate about how mathematics should be taught to engineers in the developed countries. The first is the development of mathematical modelling (MM) courses, i.e., the applications of mathematics to current engineering problems. The second is the rise of the use of micro worlds which enable student centred teaching. In this paper, we attempt to present different views and general explanation about the new development, current trends in and issues on engineering mathematics curricula, in particular *computer-based mathematical modelling*.

Keywords. Mathematical modelling, mathematics education, teaching method, microworld

1 Introduction

The pre- and continuing education of qualified and competent engineers is of great importance to every country, and mathematics plays a major role in this education whatever the levels and form of education, and the type of engineer. Based on the arising needs, there has been at least two major changes world-wide in mathematics education since 1960s. We have entered the third phase with the impact of the new information technologies (NITs), e.g., calculators, computers, interactive video, etc., on the process of teaching and learning of mathematics at all levels of schools. Yet there are numerous signs that the world-wide mathematics education is of very uneven quality, not attuned to the needs of information society, e.g., Berry [1], Blum [2].

As a result of the institutional experiences and collaboration it became apparent, on the other hand, that there is a various gap in the pre- education and continuing education of engineers in several countries, e.g., Ersoy et al [3]. This gap exists between the theoretical and practical nature of the mathematics education, i.e., between the mathematics and mathematical modelling skills. Mathematical Modelling (MM) courses, therefore, focus on the previously neglected areas of representing practical situations in mathematical form, and also require a different teaching approach and appropriate teaching/learning materials. In the present paper, we will explain our own view about the integration of the NITs, in particular microcomputers, into the mathematics curricula.

2. Mathematics Courses in the Existing System and New Trends

Mathematics is a very important aspect of Engineering. It is the principal tool used to solve problems in engineering. This is not in question. What is being questioned is the most efficient way to teach mathematics to engineers. Here we will examine the existing system and the new trends in some countries very briefly.

2.1 Teaching Mathematics in the Existing System

A typical engineering course will contain two or three mathematical modules per year, e.g., Calculus, Analytic Geometry, Linear Algebra, Differential Equation, *etc.* This will often be taught by the Mathematics department. This is because the mathematicians regard themselves as the best people to teach mathematics and believe that the engineers will cut corners and not teach the fundamental structure of the mathematics. But have we not agreed that for most engineers, mathematics is just a tool and there is often too much theory taught when it is not needed.

Suppose one had to explain how to use a spanner to a student. A series of lectures was prepared that described in detail the history and development of the spanner; its different functions and shapes; its strengths and weaknesses, its basic purpose and then a practical session where the student actually used the spanner on a series of nuts. Would this be a sensible way of teaching? Surely not. Would one not start with a practical situation where the student was given a nut which would not move under the strongest manual pressure. He was then shown how to tighten and loosen it using this tool called a spanner by applying only minimal pressure. Students who were then fascinated by how forces could be so magnified could then be taken further and taught about moments level etc. If this is the way that the use of a practical tool is taught, are there any lessons to be learned with regards the teaching of mathematics as a tool?

We would suggest that mathematics is currently taught in the first way. Students are not allowed to find rates of change unless they have done a course in calculus which involves some study of infinitesimal and some difficult symbols. This may involve a semester or even a year before the practical advantages are revealed. By this time the student is usually demotivated by mathematics and the subject has become boring, uninteresting and something to be passed to gain the degree- not an essential part of engineering harmony. Even at the final practical stage, the student is often just presented with problems and their mathematical solutions. But when left to their own devices and are faced with new problems how the students know which mathematics to apply?

The way described above is not the best way to train engineers. It creates unnatural divisions between engineering and mathematics and does not produce engineers that are resourceful, creative and flexible. What can be done?

2.2 New Trends in Teaching/Learning Engineering Mathematics

An answer to the question in the previous section is the development and introduction of MM courses. The basic idea behind such courses is to produce students who could think for themselves.

(a) Mathematical Modelling Courses: MM is, indeed, a branch of problem solving. As such it hold its own view of mathematics. To a modeller, mathematics is a self contained, consisted body of knowledge that exists in its own right. It does not belong to the real world but is a highly abstract way of thinking. Problems do exist in the real world and have to be solved. There are many ways of doing this- one being the use of mathematics.

From the Committee of SSMP Usiskin [4] emphases that there are obstacles to the implementation of applications and modelling. In many countries, including his country, USA, applications of mathematics beyond arithmetic are not part of the standard curriculum. In the presentation, he also stated, after giving some examples, that *Modelling does not appear even in the index of the books.* But good modelling courses should be open to many philosophies and solution methods. If one studies how most of the great discoveries were made one cannot fail to gasp at the distorted logic, the errors, the wrong paths and the sheer chances that led up to the theory which was later put on a secure footing by countless academics over the world.

Creativity is a human characteristic that most education systems incredibly manage to reduce as one proceeds through it. This is because one becomes accustomed to a right/wrong syndrome and tends to conform to the perceived wisdom of the day. In modelling courses, a student is never wrong. Any statement or hypothesis is merely a starting point for further investigation. Student may even finish on a different road to which they started which is perfectly acceptable. Alternative solutions and suggestions are welcomed. Students are encouraged to evaluate and to criticise their own work. In this way, it is hoped that creativity and flexibility will be encouraged to grow.

This attitude spreads to mathematics which now becomes just another tool. Students should not be taught mathematics in a MM course. No- this is the function of a mathematics course. Students should be taught that the application of mathematics to a problem is always an approximation. Sometimes the student will realise that mathematics can be applied badly. This does not denigrate mathematics, only that particular use. Students will learn why and when mathematics should or should not be used, which area of mathematics is applicable. Thus the student will be able to evaluate and criticise the use of mathematics in problem solving, i.e., problems from real world. This is a big task for any subject, and MM needs therefore new teaching methods to function efficiently.

(b) New Teaching Methods: MM must be problem and student centred. Everything must stem from the problem. The lecturer has two choices: *(i) He may suggest problems to which he has an answer; (ii) He may choose a problem at random.*

The first choice is the safest because if the student is stuck then he can be moved on. But the students just tend to reproduce the known solution and the result is similar to the existing teaching methodology described previously. The second choice is much more exiting but it has a unavoidable consequence on the student-lecturer relationship. If no-one has seen the problem before then students may suggest several solution methods. Which one is correct? The lecturer soon finds that he is no longer the found of all knowledge and will even (heaven forbid!!) make mistakes in front of the class. After a time however, the instructor will find himself more of an advisor or consultant and his/her experience and greater knowledge of mathematics will be recognised and appreciated by the students. The instructor will then gain their confidence and thus be able to help them more.

When students come to a standstill and cannot progress because of lack of knowledge, that is the time to introduce the relevant mathematics. There will be no lack of motivation or boredom now. Students will even be demanding more mathematics not less. For the problem has never been that mathematics is not useful only the fact that it has not been perceived as useful.

In this atmosphere, students will work in different ways and at different speeds. This makes the whole process student centred. In turn this means that it is very expensive in staff time and in this present time of cost efficiency is impractical. This was true till it was realised that new technologies could help the mathematics teacher and instructor. In the next section of this paper, how such courses and teaching methods are now possible due to new technologies, i.e., the computer and cognitive tools.

3 Use of Computer in Teaching Mathematics

The NITs change the nature and emphasis of the content of mathematics as well as the pedagogical strategies used to teach mathematics. Here we will explain our personal view on the use of computers in mathematics education.

3.1 New Trends in Using Technology in Maths Education

From the literature and our personal experience we understand that the use of NITs has dramatically changed the nature of physical and social sciences, business, industry and government in both industrialised and developing countries. The impact of this technological development is no longer an intellectual abstraction, but has become an economical reality, i.e., *new raw material* and *new capital.* Such changes, development and shifts should be therefore considered very seriously on time in mathematics education at all levels of schools, colleges and universities, because more people than even before are therefore seeking education in mathematical sciences in both developed and developing countries. Yet there are numerous signs that the world-wide mathematics education is of very uneven quality, not attuned to the needs of information society.

3.2 Computer-Enhanced MM Courses

The NITs change the nature and emphasis of the content of mathematics as well as the pedagogical strategies used to teach mathematics. In this connection, MM courses focus on the previously neglected areas of representing practical situations in mathematical form, and also require a different teaching approach and appropriate teaching/learning materials. There are still some constraints and certain questions which should be considered in computer-enhanced mathematics courses.

(a) Key Questions: Attempt to address the following key questions before to start designing engineering mathematics course.
* *Why do we teach mathematics and science?*
* *What should engineer learn these subjects?*
* *Can we define a "minimal level of competence" for these subjects which all engineers should reach?*
* *Would this minimal list be the same for all countries?*
* *How do we foster learners' curiosity?*
* *Where is computer use appropriate or inappropriate and why?*
* *Which topics can be more efficiently studied with computers?*
* *What is a proper balance between computer use and hand performance by traditional methods?*
* *Can computer genuinely enhance conceptual understanding of certain topics?*
* *What would be textbook and syllabus construction?*
* *How do we develop the new curricula and implement it in different situations?*

There are of course other points and questions which should be considered in the redesign of such courses.

(b) Emphases Reflected: In designing a computer-enhanced MM course, the following emphases should be reflected:
* *Focus on core topics, essential theory, methods and techniques;*
* *Concentrate on graphical and geometrical aspects;*
* *Structure the learning materials in problem solving methods;*
* *Encourage students' experimentation and exploration;*
* *Present interesting and realistic approaches;*
* *Require active in-class participation, and*
* *Demonstrate advantages of technology.*

Here we should keep in mind that the technology is changing very rapidly, and the new development, we hope, will enhance the teaching/learning environment and the process be easy as well as much better.

3.3 Microworld and Phases in the Process

One such way of using the computer is the creation of microworlds. A microworld is a set of tools pre-programmed into the computer that enable the user to tackle a problem. The recent development of Hypermedia, especially on Applemacs, enables all teachers now to construct their own micoworlds. For a modelling course, such a microwold should contain a detailed description of the modelling process and the associated pedagogy as presented to the learner on the screen. This is not intended to be a screen by screen account but rather a section by section account of the microworld contents and will include a description of the opportunities we see for learner interaction with the computer at every stage of the learning process.

In the integration of the NITs, in particular computers, into MM courses, we should always think of the following stages and or phases while teaching and learning such courses.

(a) The Exploratory Stage. In this stage of the modelling process the parameters of the model is investigated. The parameters are not looked at in depth at this stage but the learner is encouraged to vary the parameters and record impressions of the effect on the model of the parameter changes invoked.

(b) The Analytical or Identification Stage. In this stage, the learner analyses the physical processes involved and attempts to identify the important parameters of the model.

(c) The Solution Stage. Traditionally, this stage of the modelling process has been heavily dependent on the ability of the learner to cope with the problem of solving the appropriate set of equations. Essentially, the learner assumes the role of pure mathematician rather than mathematical modeller. A weaker student will use computer algebra systems (CASs) such as *Derive, Maple, Mathematica,* etc. to solve the equations resulting from the model [5-7].

(d) The Verification Stage. The learner would be expected to check that the solution of the mathematical model produced by using CASs does satisfy the basic criteria of agreeing with the known behaviour of the system. It is hoped that the learner will be sufficiently familiar with the behaviour of the system through the exploration stages of the modelling process and the expression of that behaviour in mathematical terms from the formulation stage of the modelling process that the verification stage of the learning process will not present a problem.

(e) The Predictive Stage. At this stage of the modelling process the learner would be using CASs to manipulate the model solution to predict outcomes which may not have been met or assimilated by the learner during the modelling process itself. The validity of such predictions can often be checked by running the model and looking for particular behaviour patterns.

4 Conclusions

The use of technology in the workplace and its development requires more qualified and competent engineers and scientists who have fundamental and current knowledge and some basic skills, i.e., much higher mathematics, language and reasoning capabilities. MM courses focus on the previously neglected areas of representing practical situations in mathematical form, and also require a different teaching approach and appropriate teaching/learning materials. It is our personal belief that a course in MM should form the core of the engineering mathematics curricula, and must be taught on an interdisciplinary basis and integrated to and/or part of the other relevant courses in the engineering scheme of study.

It is now possible to provide new stimulating and relevant learning environments for MM courses. Such MM courses are most effective when have been designed as an integral part of the curriculum rather than loosely attached to the existing curricula.

References

1. Berry, J. S. *et al* (eds): Mathematical Modelling Methodology, Models and Micro, Ellis Horwood, Chichester, 1986
2. Blum,W. *et al* (eds): Applications and Modelling in Learning and Teaching Mathematics, Ellis Horwood, Chichester, 1986
3. Ersoy, Y., Moscardini, A and Curran, D.: Anglo-Turkish collaboration in higher education: In: Proc. of East-West Congress of Engineering Education, Sep 1991, Cracow
4. Usiskin, Z., Building mathematics curricula with applications and modelling. In: M Niss *et al* (eds) Proc. of ICTMA-4, July 1989, Ellis Horwood, Chichester, 1990
5. Derive™ User Manual: A Mathematical Assistant for Your Personal Computer, Soft Warehouse Inc, Honolulu, 1990
6. Maple™ Reference Manual: Symbolic Computation Group, Univ. of Waterloo, Canada, 1988
7. Wolfram, S.: Matematica™: A System for Doing Mathematics by Computer, Addison-Wesley, 1992.

Examples of Matlab in Engineering Education

S. Pohjolainen [1], J. Multisilta [1] and K. Antchev [2]

[1]Tampere University of Technology (TUT), PO. Box 692, Tampere, Finland
[2]Bulgarian Academy of Sciences, ICS, Sofia, Bulgaria

Abstract. Mathematical hypermedia enables creation of mathematical virtual reality on a computer where mathematics can be studied with aid of hypertext, graphics, animation, digitised videos *etc*. In this environment the student's role can be active, he or she can make mathematical experiments and learn by doing. Numeric and symbolic computation programs play a significant role in creating mathematical hypermedia. Matlab is a programming environment for numeric computation and visualisation. It integrates numerical analysis, matrix computation, signal processing and graphics in an easy to use environment, where problems and solutions are expressed almost as they are written mathematically - without traditional low level programming. The purpose of this paper is to discuss about mathematical hypermedia and integrating Matlab as an essential part of it. As an example a hypermedia based course at TUT on matrix algebra will be discussed.

Keywords. Matlab, hypermedia, matrix algebra, tool node, hypertext, eigenvalue

1 Introduction

Computers play an essential role in the research and education of applied mathematics, natural sciences and technical sciences. During the last decade personal computers have become more and more user friendly and powerful so that nowadays many research problems can be solved with existing mathematical software. Graphical interfaces have made the use of personal computers easier. Object oriented programming environments such as HyperCard [1] and ToolBook have made it possible to easily integrate text, graphics, animations, mathematical programs and sound into hypermedia.[2-6] In the future sound and hand-writing will be included as means to control computers.

The concept *hypertext* in this context means data basis where information (text) has been organised nonsequentially [7]. The data basis consists of nodes and links between nodes. One node may contain several links to other nodes and one node may contain several links. The nodes and links form a network structure for a data basis. *Hypermedia* is a data basis, which contains in addition to text also pictures, digitised videos, animations and sound.

In hypertext a word, concept, definition or other object can be activated and the user can ask for additional information or affect the program for some other purpose. If the provided explanation is not sufficient it may be completed. Since everybody needs not to read all the material, hypertext supports reading and studying on different levels depending on the reader. Mathematical text consists typically of definitions, axioms an theorems which may be deduced from the axioms. In many courses there is quite a lot of definitions, which all should be well understood. The basic form of a mathematical hypertext is an electronic book with dictionary of definitions, which may be called and studied always when needed. The main advantage of hypertext is that it is able to adapt to the users needs. For the author this means that he has to write the text to support reading on different levels. This is always not so difficult, since hypertext is may organise the writing process, too.

Visual information is often easy to understand. The traditional, literal way of presenting mathematics is based on formulae and text. Many essential ideas in mathematics are geometrical or visual by nature and they can be shown and studied to some extent using computer graphics. With the use of mouse the graphical elements can be made interactive so that the student may travel in 3D-spaces - or even n-dimensional spaces to get better understanding on functions, surfaces, planes and so on. Animation helps understanding time dependent behaviour such as solution of ordinary differential equations, two-dimensional partial differential equations, and iteration procedures in numerical analysis, to name a few examples.

Mathematical programs such as *Maltlab* [8] in matrix algebra and numerical computations, *Mathematica* [9] and *Maple* [10] in symbolic algebra and general programs like *Mathcad* give good tools for solving problems, exercises, examples and even research problems. With these programs it is possible to solve problems of mathematical modelling in a fraction of time compared with the earlier methods where pen and paper were heavily used. One difficulty with mathematical programs has been with their syntax. In studying mathematics the student should also master the syntax of several mathematical programs. Hypermedia enables to support the student and sometimes makes it possible to transfer commands directly from the text to a mathematical program.

Mathematical modelling problems may be designed to start from the real-world outlook of the process and gradually they can be studied more and more deeply so that finally basic physical and mathematical laws controlling the process are revealed. The most important thing there is not computation but understanding the modelling process and the role of mathematics therein.

Mathematical hypermedia enables creation of a mathematical virtual reality on a computer where mathematics can be studied with aid of hypertext, graphics, animation, digitised videos etc. In this environment the student's role can be active, he or she can make numerical experiments and learn by doing.

A hypermedia based course on matrix algebra at TUT, designed according to the above principles will be discussed as an example of mathematical hypermedia [11-13] In the course of matrix algebra lecture notes, exercises and examples [13] have been written as a hypertext and links to Matlab and other mathematical programs have been made to enable numerical experiments and graphics.

2 Basics of Matlab

The basic data element in Matlab is a matrix that does not require dimensioning. This makes programming simple and allows the user to solve many numerical problems in a fraction of time as it would take to write a program in FORTRAN or C. Defining matrices is easy and elementary manipulations can be made simply by typing the formulas. Matlab contains a large number of commands of matrix algebra *e.g.* eigenvalues, eigenvectors, singular value decompositions, QR-decomposition, plotting and so on. Matlab is also a programming language where programs can be written as M-files and function subprograms. As a programming language Matlab contains

FOR-loops
WHILE-loops
IF and BREAK statements.

It is very easy to add commands and features to Matlab. Because it is able to read the position of the mouse on the screen, programs may also be interactive.

Matlab also features a family of application specific software called *toolboxes*. They form a collection of Matlab functions (M-files) that extend the Matlab environment in order to solve particular classes of problems in

* signal processing,
* control systems,
* systems identification,
* optimisation,
* neural networks,
* spline analysis,
* robust control,

and others. SIMULINK adds block diagram interface on Matlab, which is particularly useful in control systems. There is a large collection of books [14-16] , to name a few, and public domain software. Matlab is its extensible and easy to join with other programs, and it is also portable. M-files can usually directly be transferred from one computer system to another. A more compact way to transfer *e.g.* measurement data is to save it as a binary file in double precision format, which can be opened in different computer environments. The latest version of Matlab, Matlab 4, provides major improvements in sparse matrices, graphics and animation.

3 Mathematical Hypermedia

The term hypertext refers to written text, where words or other elements in the text may be selected and activated usually with a click of the mouse. Integration of hypertext with mathematical programs, videos, animations, sound and others leads to hypermedia. This way of presenting information is useful at many instances, because it is able to adapt in some degree to the users needs and instruct him in the process of learning. The structure of mathematics consists of definitions and axioms from which the theorems, corollaries and lemmas will be deduced. To understand the theorems and the essence of the courses, quite a lot of concepts should be understood. The aim of hypertext in the course of Matrix Algebra is to provide help for the student to work with these concepts so as to fully understand them.

The user interface should be made as easy as possible and in the authors opinion it should look like a book. All the buttons and icons and graphical objects should be intuitive and form a natural part of the hypertext. Fig. 1 shows a part of the hypertext course of Matrix Algebra at TUT.

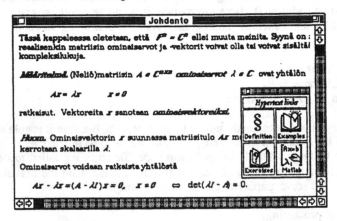

Fig.1 Hypertext lecture notes on Matrix Algebra. The user may select any word and ask for definition, exercises, examples or Matlab exercises using the menu on the lower right hand corner.

The lecture notes [5] have been written with Microsoft Word 5.0 and the mathematical formulas have been prepared with its equation editor. Then the files have been saved as Rich Text Format (RTF) [17]. This format contains all the text formatting information, pictures and formulae and it can be transferred between different word processors or between Macintoshes and PCs. Then an program, called RTF-reader [18], actually a XCMD in HyperCard was designed to read the files and to create the hypertext with mathematical formulae on a scrolling field in HyperCard. These files will then be saved in hypertext format to improve the opening speed.

Every word in the text may be selected and the user may ask for *definition*, *examples*, *exercises* and *numerical* work with Matlab. For example a user who is interested in eigenvalues can activate the word *eigenvalues* (in Finnish) and using the table of buttons in the right lower corner he can get definition, examples, numerical and theoretical exercises. Pressing the corresponding button opens a new window which contains the additional information or an exercise sheet. A scrolling window, which contains the hypertext is much easier to browse or read as a card or a page which changes instantaneously. Although the number of open windows may become large, the authors feel that it is better than opening the hypertext on the page where the definition has been given. Then it might become difficult to remember from which page the user has started from and the original sentence, which should be explained, is not seen. The definitions can be written so that they to contain more explanations than the original text. This is helpful when one wants to read hypertext e.g. in solving the exercises.

4 Structure of hypertext

A text can be divided into nodes in many ways. In Matrix Algebra course subchapters form a database, which can be read as original lecture notes on the paper. In addition all the definitions, examples and exercises are saved as files on their own nodes. The hypertext consist of definition-, example-, exercise-, tool- and card nodes. The user can study the node by selecting it with the mouse and by clicking the corresponding button.

4.1 Definition node

The definition node contains the definition of a mathematical object. The definition can be an elementary definition such as determinant, but also an end level concept like LU-decomposition. Definitions are written in as complete form as possible and they can be red as hypertext so that they may be cross referenced. In addition to a mathematical definitions, verbal explanations will also be given.

4.2 Example node

The example node gives examples on the selected topics which can be activated in hypertext. The purpose of the examples is to clarify the definitions, decompositions and methods. In the example associated with the definition of determinant, values of determinants may be computed in different ways. To present iteration methods, 3D-graphics, solutions of partial differential equations and so on, Apple Quick-Time animations may be used.

4.3 Exercise node

Exercise node gives exercises on given topics. There will be at least three kinds of exercises which may be studied on different levels.

4.3.1 Check questions

Some questions will be posed to check whether or not the student has read and understood the material sufficiently well. Questions like "Is non-singular matrix invertible" may be answered simply by pressing "yes" or "no" buttons. Some of the numerical exercises may be checked this way.

4.3.2 Numerical exercises

Numerical exercises are useful to clear out algorithms and the structure of proofs. They can be realised with an exercise sheet where the problem has been written and on which the student writes his/her solution. The program informs the user whether the proposed solution is correct or not. The correct solution can be seen by asking for it by pressing a button. This works well if the solutions are unique. Fortunately there are quite a lot unique methods and decompositions in linear algebra. For the ease of manual computation and checking the results, calculations could be restricted into the class of integer matrices. Luckily these form a sufficiently rich matrix ring where most of the methods can be demonstrated with realistic degree of difficulty. In fact, quite a lot of numerical exercises with integer matrices can be generated with computer. There are results such as: An integer matrix has an integer inverse matrices if and only if its determinant equals plus or minus one. These exercises may be selected randomly so that a student faces different problems every time the book is opened.

Another, crude method is to generate exercises, is to start from the solution. In the case of a decomposition, for example singular value decomposition, the correct U, D, V integer matrices may be selected and the matrix A to be decomposed as the product of the integer matrices A=UDV* will be an integer matrix.

4.3.3 Theoretical exercises

In the theoretical exercises a student will be asked to prove something like

> *"Show that LU-decomposition without permutation matrix for a square nonsingular matrix is unique".*

If he/she is not able to find the solution he may ask for hints. The first level of hints explains the question verbally in more detail and gives instructions about what to do. The second level of hints explains the solution. The results needed in the proof will be given in the correct order. The third level of hints gives the complete solution. Previous examination questions form a part of exercises and they should form the most motivating class of problems for a student .

4.4 Tool node

A hypermedia course should have a link to one or several existing mathematical programs such as Matlab/Simulink, Mathematica, Maple etc. In the course of Matrix Algebra a natural companion is Matlab. At this moment the easiest way is to start matlab from the text or from the exercises/examples is to use Matlab scripts (M-files), which start Matlab and run the script at the same time. A simple Matlab exercise to explain QR-algorithm can be given as follows.

The script opens Matlab and runs a QR-iteration which can be seen as a movie on the screen. Depending on the initial matrix, it will be reduced into Schur form, in which the eigenvalues of the matrix are on the diagonal.

$$
S = \begin{matrix}
2.3845 & -0.1781 & 0.6435 & -0.0209 & -0.3121 \\
0.0000 & -0.1114 & 0.8233 & -0.4086 & 0.1951 \\
-0.0000 & 0.4359 & 0.0850 & -0.3937 & -0.3078 \\
0.0000 & 0.0000 & 0.0000 & -0.0787 & -0.4509 \\
-0.0000 & -0.0000 & -0.0000 & 0.0452 & 0.0982
\end{matrix}
$$

```
function S=kuuar(A)
% Calculate eigenvalues of A by QR-method

% Move cursor to upper left corner
home;

% 30 cycles sufficient
for i:=1:30
  [Q,R] = qr(A);
  A = R*Q;
  % Try to uncomment:
  % mesh(A);

  % Cursor back to upper left corner
  home;
end;
S=A;
```

Blocks, associated with eigenvalues with almost equal modulus, may arise during the iteration and deteriorate, or slow down the speed of the algorithm so that the Schur form will not be obtained [19]. This is just the case with the above matrix, where two blocks may be seen. At this stage the student will be asked as the second exercise to modify the QR-method with shifts to overcome this drawback As a final result an improved QR-algorithm will be obtained.

With the computer it is possible to solve problems with are not accessible with pen and paper and which are important from the applications point of view. Good fields of mathematics in this respect are ordinary and partial differential equations, where it is possible to see the solutions as animations or as functions of time. 3D-graphics is one of the new application fields, which is almost impossible to study without the computer. With Matlab it is relatively easy to define a 3-dimensional object mathematically. To show it on the screen the user has to define a projection plane or subspace and the direction of the projection to define parallel projections. For perspective projections a little more complicated procedure, as shown in Fig.2, should be used. The interesting point here is that elementary 3D-graphics uses only simple matrix algebra [20] and so it seems to be an natural applications field to show the use of linear algebra.

4.5 Card node

Graphical elements, in special animations may be shown with Apple Quick Time. In the text, a Quick Time window can be seen as a picture. In fact there will be a number of pictures and they may be shown one after another or as a movie. So iteration procedures, solutions and other time varying features may be shown. Sound can be used during the animations to point out something important just when a

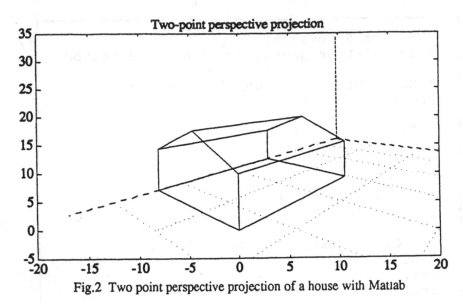

Fig.2 Two point perspective projection of a house with Matlab

person is watching live show on the screen. As the speed and capacity of personal computers increase it will be possible to record complete courses on CD-ROMs or on videodisks and use them to get the advice from the lecturer while reading the material.

5 Studying with hypertext

One of the advantages of hypertext is that it can be studied in different ways and on different levels. Learning interfaces can be seen as *orientation basis*, *contents maps*, *helps*, *hints* and supplementary questions. Orientation basis explains what should be done to solve the problem. For example an orientation map for diagonalising a matrix can be given as a flow diagram shown in Fig.3

Every box in this flow diagram can be defined to be a button. By pressing it, the user opens hypertext lecture notes on the right page or receives additional information about what to do in an new window. It is interesting that the large parts of Matrix Algebra can be studied this way.

Contents maps describe the subject matter differently from the orientation basis or from the contents of a book, which shows the sequential order of the material. Contents maps shows the inheritance of the knowledge. Fig. 4 shows contents map for singular value decomposition of a general matrix.

The above contents map is useful for a person who wants to study singular value decomposition directly without reading other parts of the text. While reading the chapter on SVD he has to understand and elaborate with the concepts on the right hand side. Hypertext enables him to study all the concepts and definitions just in the order where he finds them in the text. While this order may not be the best from the mathematical point of view, it might be the most motivating for the student.

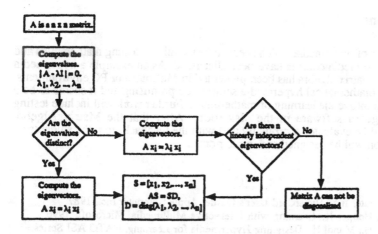

Fig. 3 Orientation basis for diagonalisation of a matrix

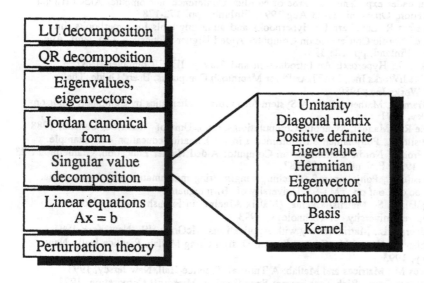

Fig.4 Contents map for singular value decomposition

5 Conclusions

Different aspects of hypermedia which seem to be useful in teaching mathematics and in designing hypermedia courses have been discussed. As an example a hypermedia based course an matrix algebra has been presented in Macintosh or PC environments. Experiences of mathematical hypermedia seem to be promising and it seems to be a natural way to improve the learning of mathematics. Further work will include testing the Matrix Algebra software in the classroom. After that the Matrix Algebra hypercourse will be evaluated thoroughly and with the feedback from the students an improved version will be designed and implemented.

References

1. Apple Computer Inc., HyperCard User's Guide, Apple Computer Inc., 1987
2. Ambron S., Hooper K.: Learning with Interactive Multimedia. Microsoft, 1990
3. Jonassen D. H., Mandl H.: Designing Hypermedia for Learning, NATO ASI Series, Vol. F 67, Springer-Verlag, Berlin, 1990
4. Jonassen D.H., G. R. Scott: Problems and Issues in Designing Hypertext/ Hypermedia for Learning. Designing Hypermedia for Learning, NAT O ASI Series, Vol. F 67, Springer-Verlag, Berlin, 1990, pp. 3-25
5. Kalaja M., et al.: Implementing an authoring tool for educational software - the Hyperreader experience. In: Proc of Nordic Conference on Computer Aided Higher Education, Otaniemi, 21-23 Aug 1991, Finland, pp. 172-178
6. Krönefors R, Lundberg L.: Hyperbooks and authoring tools for hyperbooks. In: Proc of Nordic Conference on Computer Aided Higher Education, Otaniemi, Aug 1991, Finland, pp. 188-198
7. Conklin J.: Hypertext: An Introduction and Survey, IEEE Computer, 1987
8. The MathWorks Inc., MATLAB™ for Macintosh Computers. Users' Guide, The MathWorks Inc., 1989
9. Wolfram S., Mathematica™ A System for Doing Mathematics by Computer, Addison-Wesley, 1991
10. Maple Ref Manual, Symbolic Computation Group, Univ.of Waterloo, Canada, 1988
11. Multisilta J., Pohjolainen S.: Multimedia in mathematics education - an example. In: Proc of Nordic Conference on Computer Aided Higher Education, Otaniemi, 21-23 Aug. 1991, Finland, pp. 179-187
12. Multisilta J., Pohjolainen S.:Teaching engineering mathematics with hypermedia. In Proc of Conf on TMT-93, University of Birmingham,
13. Pohjolainen S.: Matriisilaskenta (Matrix Algebra, in Finnish). Lecture Notes, Tampere University of Technology, 1993
14. Goldberg J.L.: Matrix Theory with Applications, McGraw-Hill, New York 1991
15. Shahian B., Hassul M.: Control System Design Using Matlab, Prentice-Hall, New Jersey, 1993
16. Marcus M.: Matrices and Matlab: A Tutorial, Prentice-Hall, New Jersey, 1993
17. Microsoft Corp., Rich Text Format Specification. Microsoft Corporation, 1992
18. Multisilta J.: XCMD for hypermedia applications. MacTech Magazine (to appear)
19. Ciarlet P.G.: Introduction to Numerical Linear Algebra and Optimisation. Cambridge University Press, Cambridge, 1989
20. Foley J.D. et al: Computer Graphics: Principles and Practice, Addison- Wesley, 1991.

Issues Involved in Teaching Calculus with Computer Algebra Systems

F. Simons
Technical University of Eindhoven, Eindhoven, The Netherlands

Abstract. The availability of mathematical software not only for users of mathematics but also for students has many consequences for the teaching of services courses in mathematics. Based on many years experience, the author presents his personal view on what should or could be done.

Keywords. Calculus, mathematical software, computer algebra, equation, line integral, Newton's method, implicit function

1 Introduction

Mathematics is increasingly used in many areas, and therefore service courses in mathematics to non-mathematicians can be found at nearly all universities. It is very remarkable that all over the world these courses are very similar and consist of some calculus and some linear algebra.

The contents of these courses dates from the times that all mathematical computations had to be done by hand with pencil and paper, and in fact they reflect the needs of scientists and engineers of decade ago. Nowadays a computer is an indispensable tool for everyone who use mathematics professionally and the software that can be used ranges from simple numerical and graphical packages to very sophisticated computer algebra packages. This strongly influences the way mathematics is used nowadays and in turn this should have an impact on the contents of the service courses.

My personal concern in the use of computers in the service teaching of mathematics already dates from many years ago. When simple pocket calculators became available, I started using them with the students for example in finding the sum of a convergent series. After all, that is what series is about. However, in most courses on calculus quite a lot of attention is paid in demonstrating that some series have a sum, but apart from very trivial or artificial cases rarely an attempt is done in finding the value of the sum.

Soon the calculators were replaced by programmable pocket computers, on which the students implemented numerical procedures, e.g. for finding zeros of a function of one variable and for numerical integration. This had strong impacts on the course. As an example, consider the integrals

$$\int_0^1 x\,dx, \quad \int_0^{\pi/2} x \arctan x\,dx, \quad \int_0^1 x^5 \sin x\,dx.$$

Traditionally, the first integral is solved by heart, the second through integration by parts and the last with a reduction formula. Now the blame must be laid on the student doing the first integral by a computer, but there is no convincing answer to a student who claims that the computer can find the values of the second and third integral in a shorter time than the instructor. At that time (1983) a pure mathematician might have answered that he wants to see the exact answer instead of a numerical approximation, but that argument is not longer valid. Now any computer algebra package returns the exact value of these integrals in less than a few seconds. Hence the introduction of numerical methods for finding integrals forced us to reconsider the relevance of analytical integration techniques in the course.

The next step was the replacement of the pocket computer by the personal computer, thereby obtaining powerful graphical facilities and the possibility of using more powerful numerical techniques. Again this had a strong impact on the contents of the course. For example, since a computer can easily plot the graph of a function and find the zeros and extreme values of that function numerically, also in cases where analytical solutions are impossible, we had to decide how important the techniques of the traditional courses (using the sign of the derivative) really are and to find a balance between the graphical, numerical and analytical techniques available.

The availablity of graphical and numerical facilities also changed the didactical presentation of many topics. For example, when introducing the concept of a Riemann integral in the very beginning one has to discuss the Riemann sum. Such a Riemann sum can be computed numerically and therefore it is very natural to start with the numerical evaluation of Riemann integrals before turning to analytical methods.

Differential equations are another important example. A first order differential equation can be considered as a direction field. With a computer it is easy to plot a direction field and students have no problems in sketching solutions in a direction field. At this stage I ask the students why they sketch the graph in the way they do it; they discover themselves Euler's method. Hence I start with a discussion on this method and some of its improvements. After the first lecture on differential equations the students can solve practically all first order differential equations numerically. That some differential equations can also be solved analytically is shown later on.

The latest development in our course is the introduction of computer algebra packages. Some years ago we started using DERIVE, and now we are experimenting with MATHEMATICA. It is likely that starting in 1994 we will use MATHEMATICA for the whole population of first year mechanical engineering students. Probably due to the fact that we had adapted our courses already to numerical and graphical software, the introduction of computer algebra did not have a very strong impact on the contents of our course. As an example, one of the consequences was that we decided not to teach analytical techniques for finding primitives or definite integrals any more. When a not too elementary integral is met, the students may ask the computer. At first sight this may sound very revolutionary, but for us it was only a small step. The importance of analytical techniques for finding integrals had already strongly diminished when we introduced numerical techniques, and with computer algebra there is just no need for knowing these techniques.

However, using computer algebra did have an important consequence: it changed the way we are doing mathematics ourselves. We now do time consuming symbolic computations with the computer instead of with pencil and paper and found that some techniques that are important for hand computations are unpractical for computer algebra and therefore had to be replaced by different techniques that would not have been chosen for hand techniques. This experience also had its impact on the contents and the didactics of the course.

2 The fundaments of a calculus course with computer algebra

Nowadays there are many discussions on how computers should be used in the teaching of mathematics. Many conferences are held, many journals exist on technology in mathematics teaching. It strikes me that most discussions are in the field of computer aided teaching: how can a computer help the teacher to teach what he is already teaching for many years? Of course a computer can be very useful for demonstrations (though I think that I can explain the concept of a Riemann integral better with a piece of chalk and a blackboard than a computer can do it), but for me this is not the topic we should concentrate on. With a computer we can solve many problems that are out of the range of a traditional calculus course either analytically or numerically. Hence for me it is evident that what we should concentrate on is developing completely new courses in calculus, realizing that the students will use later on in their professional career computer software, in particular computer algebra, for solving the mathematical problems they will meet. Hence students must get acquainted with such packages already at an early stage, that is when they study the course, and the software should be available throughout the course.

Most calculus courses concentrate on exercises. Many of these can now be solved easier and much faster by using computer algebra than by hand. Introduction of computer algebra in a calculus course has the danger that students are going to concentrate on which keys have to be pressed instead of on the mathematics. Even when a problem can not be solved straightforward by a computer algebra package, it is possible to extend the package such that also more complicated problems can be solved by pressing keys. In the next section I will present some examples. Hence we have to be very careful what we really want to teach to the students: solving standard problems by pressing keys or understanding of the mathematical concepts. This is a choice that has to be made. Teaching concepts is much more difficult (for the students and for the teacher) than teaching how to solve exercises.

Most of the concepts that are taught in traditional calculus courses have turned out to be and still are very important. In a new calculus course the concepts will be roughly the same, but the way they are taught will strongly differ, particularly in the way they are applied when a computer algebra package is available. Further, the use of a computer algebra package can make it necessary or allows to include other topics in the course as well.

I feel slightly disappointed on the changes that are made so far. Many of the new books on calculus, in which modern software such as DERIVE, MAPLE or MATHEMATICA is used, concentrate on which keys have to be pressed for finding the desired result. Using the software in this way does hardly contribute to a better mathematical understanding. For example, they tell you how to plot the graph of the function

$$f(x) = \sqrt{1 + x^4} - 3x$$

and how to find the zeros and the minimum, but they do not pay attention to the fact that the graph of this function can and should be sketched by heart before switching on the computer. It is the sum of the almost trivial functions $\sqrt{1 + x^4}$ and $-3x$; for large x the function appears like $x^2 - 3x$ and for x close to zero the function looks like $1 - 3x$. It is a matter of attitude first to predict what the computer will show and then use the computer only for verification and finding details such as zeros and extreme values.

A starting point for developing a new course with computer algebra could be that what a computer can do better than we can do it ourselves should be done by a computer. The small computer algebra packages, though already remarkable powerful, are often too small for 'real-life' problems. On the other hand, everyone who uses more advanced computer algebra packages will have the experience that for an efficient use of these packages one needs a thorough understanding both of mathematics and of the package. The user must know how the problem could be solved mathematically and moreover how an efficient use can be made of the computer algebra package for doing the computations. Hence if the aim of the course is to prepare the students to be able to work in this way, the level of the course will automatically be higher than that of a traditional course. The accent will be more on mathematical insight than on how to solve a certain problem. Of course students must have a certain skill in solving problems, but it is important that realize why they solve the problem in the way they do it rather than that they concentrate on how we say they have to solve it. Mimicking gimmicks does not contribute to a better mathematical understanding, no matter whether it concerns analytical techniques or computer commands.

The general experience of people who try to use computers in their calculus teaching is that one continuously has to realize what one really wants to teach. Why do we include this topic, what should the students know about it, on what level? What analytical techniques will we teach, what can the computer do? The introduction of software in the course strongly confrontates the teacher with the aims of the course; many decisions on what and how have to be taken.

For our courses we formulated the following starting points [1]:

1. The courses are courses in mathematics. That is, we concentrate on the mathematics and not on the computer or on the software packages. This sounds simple, but what is mathematics in a service course? In our opinion, the aim of a service course in mathematics is to provide the students with the mathematical tools and background that they need in their discipline, in our case mechanical engineering. Hence the contents of our courses are determined in consultation with the faculty of mechanical engineering. A thorough understanding of the tools an engineer will be using is considered to be essential and therefore a strong accent must be placed on insight. This should not be confused with mathematical rigor. In many cases, we can argue by simple reasoning why a certain property must hold without presenting a complete mathematical proof. Proofs of properties are only given when they contribute to insight. In this sense our courses differ from those for mathematics majors. We try to present a more physical way of working with mathematics, trying to make the students think before applying an analytical or computer technique, forcing them beforehand to formulate what they expect the result will be or will look like.

2. For solving a problem, students may always choose which technique they want to apply. If they think a computer technique will be simpler than an analytical technique, they always are allowed to use that technique. As a consequence, we will not give exercises of the type 'Do this by hand' or 'Do this by computer'.

3. The software we will use has to be software that the student will or might use in his professional life. The courses of course have to be independent of the software; the software is only a tool.

4. No applications from other disciplines are included in the courses. This is because our task is to provide the students with the necessary background and understanding of mathematics. They will see plenty of applications in their own field and mathematicians often do not have the background to explain the applications properly.

I should stress that this is how we feel about it and that in other situations different courses, based on different starting points, may be developed.

3 Examples

In this section some examples will be presented demonstrating some of the effects of introducing computer algebra packages in a calculus course. These examples are choosen such that they show what problems might be met or how teaching might change. The first two examples deal with equations; the third example will show that computing a line integral with a computer algebra package can be somewhat complicated, but once it is understood the computer algebra package can be extended such that computing line integrals is reduced to pressing keys. The last example deals with implicit functions. When making full use of computer algebra, one needs some mathematics that is normally not included in a traditional calculus course. At the end it turns out that making exercises again can be reduced to pressing keys.

The computations were done by the computer algebra package MATHEMATICA 2.0 on a NeXT; the figures have been drawn by this program as well.

3.1 A simple equation

In any calculus course equations must be solved. Since equations that can be solved analytically are relatively rare, we normally have to use a numerical procedure, e.g. Newton's method for solving an equation of the form

$$f(x) = 0.$$

Such procedures require an initial value that has to be chosen close to the solution we want to find. Hence before applying such a procedure we must have already an idea where to find the solutions. Often we plot the function on the left hand side, choose an initial value for each of the solutions we see in the plot and approximate each of these solutions.

Now the problem arises how we can be sure that in this way we really have found all solutions. If we plot the left hand side on a small interval, we might overlook solutions for values of the variable outside this interval and if we plot on a large interval, the plot may be scaled in such a way that we overlook solutions close to each other. Hence many mathematicians require that the student also proves that all solutions are found. In some cases, e.g. polynomial equations of not too high degree, this indeed can be done but in general I think it is not realistic to ask for a proof that all solutions have been found.

As an example, consider the equation

$$\sin(x) = \arctan(x).$$

How many solutions has this equation? Obviously, $x = 0$ is a solution. Many people think that since $\arctan(x)$ tends to $\pi/2$ as $x \to \infty$ and $\sin(x)$ is bounded by 1, this is the only solution. However, if we look at the third order approximations at $x = 0$ of $\sin(x)$ and $\arctan(x)$ we conclude that for small positive x we have $\sin(x) < \arctan(x)$. This is confirmed by the plot in figure 1. For positive x there is at least one solution. This solution can be approximated:

```
In[1] := FindRoot[ Sin[x] == ArcTan[x], {x, 1.5} ]

Out[1] = {x -> 1.55709}

In[2] := {Sin[x], ArcTan[x]} /. %

Out[2] = {0.999906, 0.999906}
```

Due to symmetry, of course $x = -1.55709$ is another solution.

So we have found three solutions of the equation. How can we be sure that there are no others? The plot in figure 1 suggests that there are no other solutions, but

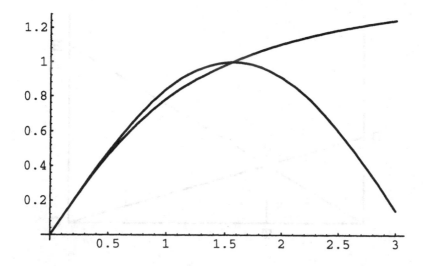

Figure 1 Graphs of sin and arctan

can you really find a *proof* that there are no other solutions? And if so, do you think your students should be able to find this proof?

3.2 The ladder problem

The ladder AB has length 4, the ladder CD has length 3 and the point of intersection S of the ladders is at height 1 above a street, as shown in figure 2. The problem is to find the width x of the street.

This well-known problem can be solved in many ways, but none of them is obvious for the students. Probably the simplest is to consider the length of ST as a function $h(x)$ and solve the equation $h(x) = 1$.

Using a computer algebra package, polynomial equations are usually easier to handle than equations involving transcendental functions. Hence we can also try to find a set of polynomial equations. But then the problem arises which variables other than x we have to use. One might start with putting $AD = y, CB = z$ and find equations for x, y and z. But even with a computer algebra package it is important to introduce not too many variables. In this problem one extra variable will do. We define

$$p = AT/AC.$$

Then

$$AT = px, \quad TC = (1 - p)x, \quad AS = 4p, \quad SB = 4(1 - p), \quad DS = 3p,$$

$$SC = 3(1 - p).$$

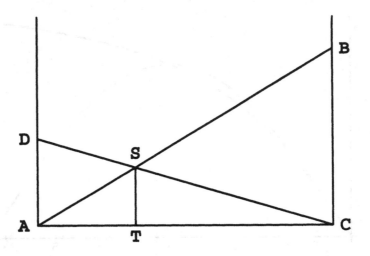

Figure 2 The ladder problem

Applying Pythagoras' theorem to the triangles ATS and CTS yields two equations:

```
In[1]  := eq1 = p^2 x^2 + 1 - 16 p^2 == 0

                 2     2 2
Out[1] = 1 - 16 p  + p  x   == 0

In[2]  := eq2 = Expand[ (1-p)^2 x^2 + 1 - 9 (1-p)^2 ] == 0

                          2     2         2     2 2
Out[2] = -8 + 18 p - 9 p  + x  - 2 p x  + p  x   == 0
```

Now every student knows what to do: the variable p has to be eliminated and from the resulting equation we need the solution x between 0 and 3.

```
In[3]  := Eliminate[ {eq1, eq2}, p ]

                 2        4        6     8
Out[3] = -5374 x  + 763 x  - 46 x  + x   == -13585

In[4]  := NSolve[ %, x ]

Out[4] = {{x -> -3.92241 - 0.0721765 I},

   {x -> -3.92241 + 0.0721765 I}, {x -> -2.90907},

   {x -> -2.60329}, {x -> 2.60329}, {x -> 2.90907},
```

```
{x -> 3.92241 - 0.0721765 I},

{x -> 3.92241 + 0.0721765 I}}

In[5] := Select[ x /. %, Positive]

Out[5] = {2.60329, 2.90907}
```

Hence we find two solutions for the ladder problem. On the other hand, if we consider the function $h(x)$ defined as the length of ST as a function of x, then it is clear that this function is decreasing for $0 < x < 3$, so the equation $h(x) = 1$ can have at most one solution. What went wrong?

Actually nothing went wrong. One has to realize what elimination really means. In our situation it gives an equation in x such that for every solution for x of this equation our original set of equations has a solution for p. So what we have to do is to find for each of our two x-values the corresponding p-value and choose that value of x for which the corresponding value of p is between 0 and 1. Now a minor problem occurs. If we substitute the x-value in the two equations, the resulting two equations in p will have no solutions, since we substituted only an approximate x-value. Hence we have to start from scratch, first solve the equations for p and then find the value of x for that value of p between 0 and 1. This is best performed by looking at the Groebner basis:

```
In[6] := gb = GroebnerBasis[ {eq1[[1]], eq2[[1]]}, {x, p} ]

                       2        3        4
Out[6] = {-1 + 2 p + 7 p  - 14 p  + 7 p ,

       2       3    2
  -5 - 21 p + 14 p  + x }
```

From this basis we conclude that p has to be the zero between 0 and 1 of the first polynomial and that by the second polynomial we have an explicit formula for x as a function of p:

```
In[7] := Solve[ gb[[2]]==0, x ]

                    2       3
Out[7] = {{x -> Sqrt[5 + 21 p  - 14 p ]},

                   2       3
  {x -> -Sqrt[5 + 21 p  - 14 p ]}}

In[8] := x = x /. %[[1]]

                 2       3
Out[8] = Sqrt[5 + 21 p  - 14 p ]
```

```
In[9] := NSolve[ gb[[1]] == 0, p ]
```

```
Out[9] = {{p -> -0.364244}, {p -> 0.329281},
```

```
  {p -> 1.01748 - 0.394738 I}, {p -> 1.01748 + 0.394738 I}}
```

```
In[10] := x /. %[[2]]
```

```
Out[10] = 2.60329
```

So this is the width of the street.

This example shows that eliminating can be easily done by a computer, but that we have to understand what eliminating is. Do we also have to teach Groebner bases?

If we had solved the two equations $eq1$ and $eq2$ for x with MAPLE, the result would be presented (by using Groebner bases) in a similar form as Out[7] and the remark that p has to be a zero of the first polynomial in Out[6].

3.3 Line integrals

A topic contained in most calculus courses is that of the line integral. Let a function f of two or three variables be given and let a curve in the coordinate space be given by a parametrization. Finding the integral of the function over the curve requires some computational skill and at the same time an understanding what has to be done. We differentiate the parametrization to the parameter and compute the length of the resulting vector; we substitute the parametrization in the function, multiply both results and integrate between the limits for the parameter. Usually this integral can be evaluated only numerically. For computations like these a computer algebra package can be very helpful. As a demonstration we compute

$$\int_C f \, ds$$

with $f(x, y, z) = x^2 + 3yz$ and $C = (t, t^2, t^3)$, $0 \le t \le 1$. The command Norm that we will use is one of the commands that we have added to MATHEMATICA for our calculus course.

```
In[1] := f[x_,y_,z_] := x^2 + 3 y z
```

```
In[2] := par = {t, t^2, t^3}
```

```
              2   3
Out[2] = {t, t , t }
```

```
In[3] := v = D[ par, t ]
```

```
                    2
Out[3] = {1, 2 t, 3 t }

In[4] := (f @@ par) Norm[ v ]

              2     4   2     5
Out[4] = Sqrt[1 + 4 t + 9 t ] (t  + 3 t )

In[5] := NIntegrate[ %, {t, 0, 1} ]

Out[5] = 2.37102
```

This example shows that one has to know the algorithm how to compute a line integral; the only thing the computer does is the computations. The example also shows that one must be very familiar with the package to be able to perform the operations. We defined f as a function and not as an expression to be able to substitute the parametrization by means of an Apply command.

One of the things a clever student or the teacher can do is putting this algorithm in a new command, for example in the following way:

```
In[6] := NLineIntegral[
             exp_, par_, {t_, a_, b_}, opts___ ] :=
  Module[ {v},
    v = Sqrt[ Plus @@ (#^2&) /@ D[ #[[2]]& /@ par, t ] ];
    NIntegrate[ (exp /. par) v, {t, a, b}, opts ]
    ]
```

This is a very short program and the computation of line integrals now has been reduced to pressing keys:

```
In[7] := NLineIntegral[ x^2 + 3 y z,
                       {x->t, y->t^2, z->t^3}, {t, 0, 1} ]

Out[7] = 2.37102
```

Is this the way we want to use a computer algebra package in a calculus course?

3.4 Implicit functions

Most calculus courses pay some, but not much, attention to implicit functions. We consider a curve given by an equation $f(x, y) = 0$. Locally we can consider a part of the curve as the graph of function $y(x)$. As an example we consider the curve

$$x^5 - x^3 y + 3y^4 - 3y^2 + xy = 0.$$

With MATHEMATICA we can get a rough idea of the curve by using the command ContourPlot. The result is shown in figure 3 at the left hand side.

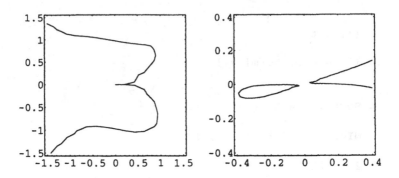

Figure 3 The curve $x^5 - x^3y + 3y^4 - 3y^2 + xy = 0$

The figure suggests that the curve has a cuspidal point at (0,0). However, by careful inspection of the formula of the curve, it is not too difficult to get an idea of the behaviour of the curve near the origin. The terms of degree 3 and higher can be neglected for x and y close to zero with respect to the second order terms. That means that the curve near the origin appears like

$$xy - 3y^2 = 0.$$

Hence the origin is a double point, and the two tangent lines are $y = 0$ and $y = x/3$. If we make a plot closer to the origin and with the number of plotpoints increased to 40, we get the right-hand side of figure 3. It is much better than the left hand side, but still not in accordance with reality.

The curve passes through the point (0,1) and by the implicit function theorem, a function $y(x)$ exists such that $f(x, y(x)) = 0$ and $y(0) = 1$. The derivatives of $y(x)$ at $x = 0$ are usually found by hand by implicit differentiation. That technique can be done with MATHEMATICA as well, but not very comfortable:

```
In[1] := exp = x^5 - x^3 y + 3 y^4 - 3 y^2 + x y

             5         3       2       4
Out[1] = x   + x y  -  x  y - 3 y  + 3 y

In[2] := y[0] = 1

Out[2] = 1

In[3] := exp == 0 /. y->y[x]

             5           3          2         4
Out[3] = x   + x y[x] -  x  y[x] - 3 y[x]  + 3 y[x]    == 0
```

```
In[4] := D[%, x]

            4              2                      3
Out[4] = 5 x  + y[x] - 3 x  y[x] + x y'[x] - x  y'[x] -

                       3
     6 y[x] y'[x] + 12 y[x]  y'[x] == 0

In[5] := % /. x->0

Out[5] = 1 + 6 y'[0] == 0

In[6] := y'[0] = y'[0] /. Solve[ %, y'[0] ] [[1]]

           1
Out[6] = -(-)
           6

In[7] := D[ %%%, x ]

            3                         2                2
Out[7] = 20 x  - 6 x y[x] + 2 y'[x] - 6 x  y'[x] - 6 y'[x]  +

         2     2                3
     36 y[x]  y'[x]  + x y''[x] - x  y''[x] -

                         3
     6 y[x] y''[x] + 12 y[x]  y''[x] == 0

In[8] := % /. x->0

         1
Out[8] = - + 6 y''[0] == 0
         2

In[9] := y''[0] = y''[0] /. Solve[ %, y''[0] ] [[1]]

            1
Out[9] = -(--)
           12
```

The main reason why we are interested in these derivatives is that they enable us to form the Taylor polynomial of the function $y(x)$ at $x = 0$. This Taylor polynomial, and thereby the derivatives, can be found much easier with a computer algebra package in the following way:

```
In[10] := pol = 1 + Sum[ a[n] x^n, {n, 1, 5} ] + O[x]^6
```

```
                              2         3         4
Out[10] = 1 + a[1] x + a[2] x  + a[3] x  + a[4] x  +

      5       6
   a[5] x   +O[x]
```

```
In[11] := Solve[exp == 0 /. y->pol ]
```

```
                      19            253           4
Out[11] = {{a[5] -> -(---), a[4] -> ----, a[3] -> --,
                      162           3456          27

         1               1
   a[2] -> -(--), a[1] -> -(-)}}
         24              6
```

```
In[12] := pol /. %[[1]]
```

```
               2     3        4        5
          x   x    4 x    253 x    19 x           6
Out[12] = 1 - - - - -- + ---- + ------ - ----- + O[x]
          6   24   27     3456    162
```

However, if we need higher order approximations, a much better technique exists. In fact it is Newton's method for approximating a zero of a function of one variable. If we have to solve the equation

$$g(y) = 0,$$

Newton's method uses iteration with an initial value close to the zero with the iteration function

$$it(y) = y - \frac{g(y)}{g'(y)}.$$

To solve $y(x)$ from an equation $f(x, y(x)) = 0$ the same method can be applied, the only difference being that now a parameter x occurs in the iteration function.
In the example we construct the iteration function with MATHEMATICA:

```
In[13] := it = Function[ y,
                Evaluate[Together[y - exp/D[exp, y]]]]
```

```
                        5     2     4
                      -x  - 3 y + 9 y
Out[13] = Function[y, -------------------]
                        3         3
                      x - x  - 6 y + 12 y
```

For x close to 0, a good initial value will be the function $y_0(x) = 1$. However, iteration with functions soon results in very complicated expressions, which will be clear if we iterate only three times. Hence we need some more theory.

It is well known that Newton's method converges quadratically for solving an equation numerically. A similar property holds when we are working with functions instead of numbers: if $y(x)$ is a solution of $f(x, y(x)) = 0$ and if

$$y(x) = p(x) + O(x - a)^n,$$

then

$$y(x) = it(p(x)) + O(x - a)^{2n}.$$

Since in our example we indeed have

$$y(x) = 1 + O(x),$$

it is easy to find to find high-order approximations for $y(x)$. In the following output the cryptical MapAt statement is used to double the exponent of the O-symbol.

```
In[14]  := 1 + O[x]

Out[14] = 1 + O[x]

In[15]  := it[ MapAt[ 2#&, %, 5 ] ]

                x    2
Out[15] = 1 -  - -  + O[x]
                6

In[16]  := it[ MapAt[ 2#&, %, 5 ] ]

                   2      3
              x   x    4 x         4
Out[16] = 1 - -  - --  + ----  + O[x]
              6   24      27

In[17]  := it[ MapAt[ 2#&, %, 5 ] ]

                   2      3        4         5           6
              x   x    4 x    253 x     19 x     28937 x
Out[17] = 1 - -  - --  + ---- + ------  - -----  - -------- -
              6   24      27     3456      162       248832

       7
  1265 x            8
  ------- + O[x]
   11664
```

```
In[18] := it[ MapAt[ 2#&, %, 5 ] ]

                   2     3      4        5          6
              x    x    4 x    253 x    19 x    28937 x
Out[18] = 1 - - -  -- + ---- + ------ - ----- - -------- -
              6   24    27     3456     162      248832


        7             8             9               10
  1265 x       258515 x       445237 x       371186633 x
  ------- +   --------- +   --------- +   ------------- -
   11664       23887872       5038848       5159780352


         11                  12                   13
  1699549 x         93672366319 x         4689313921 x
  ----------- -    ---------------- -    -------------- -
   120932352        743008370688          34828517376


               14                   15
  609449883731 x       275498764865 x                16
  --------------- +    --------------- + O[x]
   17832200896512       2507653251072
```

This example shows that when using computer algebra the so called power series substitution is easier to use than implicit differentiation for finding Taylor approximations of implicitly defined functions and that Newton's method is even more powerful. The result is a fast algorithm for finding Taylor polynomials of implicitly defined functions. Hence we can extend MATHEMATICA with a function for finding Taylorpolynomials of implicitly defined functions in the following way:

```
In[19] := ImplicitSeries[
    eq_Equal, {x_, a_, n_}, {y_, b_} ] := Module[
  { exp, test, p },
  exp = eq[[1]] - eq[[2]];
  test = N[ exp /. { x->a, y->b } ];
  p = If[ test === 0, b,
          y /. FindRoot[ exp /. x->a, {y, b} ] ] + O[ x, a];
  it = Together[ y - exp / D[ exp, y ] ];
  While[ p[[5]] < (n+1),
    p[[5]] = Min[ 2 p[[5]], n+1];
    p = it /. y->p
    ];
  p]
```

This function ImplicitSeries is called with three arguments. The first one is the equation, the second one a list consisting of the independent variable, the value around which we want to find the Taylor polynomial and the order and the last one a list containing of the dependent variable and the corresponding value. If this value does not satisfy the equation it is considered as the initial value for solving the

equation and the result will be a numerical approximation of the Taylor polynomial.
Here is how it works:

```
In[20] := ImplicitSeries[ exp == 0, {x, 0, 5}, {y, 1} ]
```

$$\text{Out}[20] = 1 - \frac{x}{6} - \frac{x^2}{24} - \frac{4\,x^3}{27} + \frac{253\,x^4}{3456} - \frac{19\,x^5}{162} + O[x]^6$$

```
In[21] := ImplicitSeries[ exp == 0, {x, 0.5, 5}, {y, 1} ]
```

$$\text{Out}[21] = 0.923249 - 0.126969\,(-0.5 + x) -$$

$$0.00756399\,(-0.5 + x)^2 - 0.412784\,(-0.5 + x)^3 -$$

$$0.894197\,(-0.5 + x)^4 - 0.949209\,(-0.5 + x)^5 +$$

$$O[-0.5 + x]^6$$

Again we have developed the theory so far that we can solve the problems with pressing keys. What do we want to teach to the students?

4 Some conclusions

It seems to be evident that the traditional calculus courses have to change. One reason is the fact that computer algebra starts getting available both for users of mathematics and for the students. Another reason is that the results of the nowadays calculus courses seem to be unsatisfactory. A fairly common complaint is that although the students have passed the calculus examinations, they are not able to do anything with it. Some literature even uses the term calculus crisis.

In an attempt to reform the calculus teaching at least three parties are involved.

- the professional users of mathematics, in particular the department we are teaching for. For them the use of computers in general and of computer algebra in particular might have changed the way they use mathematics nowadays and therefore it is important that developers of new courses get feedback from them what mathematics on what level is needed. This sounds trivial, but it is not. Many users of mathematics still have no idea what modern software can do for them and how it changes the way they will apply mathematics. Further, what users can do with mathematics strongly depends on his or her level of understanding of mathematics. For some, mathematics is a box of gimmicks that can be applied when needed, for others it is the basis of their daily work. Therefore some will argue that the amount of mathematics can be reduced, particularly since the computer now can perform the tricks for them. Others will ask for courses with accents on concepts and mathematical reasoning at a higher level than nowadays.

- the teachers of the future mathematics courses. The traditional calculus can be done easier and simpler with computer algebra, so what has to be done with the course? Stick to the old contents and solve the traditional problems with pressing keys? Or invent new courses playing with the new tools and discovering that mathematics, even on the so called calculus level, is inspiring, challenging, non-standard and puts a higher demand on understanding? Anyway, it requires a reconsideration what is considered to be important and what is less important or maybe even irrelevant.

- the students. The good students will find a course with computer algebra more interesting and more challenging than a traditional course. On the other hand the bad students get completely lost, at least to our experience. Since we introduced computer algebra, the failure for our examinations has increased. This is not surprising: in a traditional course a student can pass an examination by simply learning some standard techniques by heart without understanding. If computer algebra is available, such examination questions will not longer be asked.

Computer algebra as such does not contribute to a better understanding. On the contrary, it requires a better understanding in order to be able to make full use of it. So here is a dilemma: if we try to increase the understanding in our math courses, less students will pass. On the other hand, there certainly is a demand for scientist and engineers with a good understanding of mathematics.

The solution seems obvious: we will have to present courses on several levels. Maybe in the future for most students a calculus course will concentrate on how to solve elementary problems by using only some elementary commands of a simple computer algebra package. For the mathematically more gifted students courses could be given that concentrate more on insight than on techniques. Unfortunately, this is a political issue. The idea that with respect to mathematics some people are more gifted than others is for many politicians difficult to accept.

Finally, the teaching process itself can be a barrier for emphasis on understanding. Partly this is due to the attitude of the majority of the students entering our university. They are not interested in mathematics as such, but only in how to pass the examinations. For that they only want to know how they should solve the examination problems. Also, we have to teach large groups of students. Hence many people are involved in the course and that requires a certain standardization of the material, in particular with respect to the exercises that are discussed during the tutorials. Hence it seems unavoidable that also our course is going to concentrate on a limited set of exercises, to be solved with a set of standard techniques (different from those of a traditional calculus course), instead of a varied collection of exercises on which the students can test their insight and inventivity. I think this danger is less when the need of standardization is not present.

References

1.   J.J.M. Rijpkema, F.H. Simons, J.G.M.M. Smits: Mathematics courses with a PC. *Int. J. Math. Educ. Sci. Technol.* **22** (1991), 791-798

References

A Global Approach to Finite Element Method

M. M. Cecchi and E. Secco
Dipt di Matematica Pura ed Applicata, Università di Padova, Italy

Abstract. A global and unified approach to teach finite element presented here is more convenient to introduce most applications for problems in stress analysis as well for problems in fluid dynamics and in geomechanics as problems in elasto-plasticity. The approach is also useful for many very different topics, because the common theoretical background helps to give a convenient and unique presentation in several applications. As examples of applications a study of a shallow water problem and a very unusual application of stability study into a quarry are presented.

Keywords. Global approach, finite element method, shallow water problem, tidal flow, Galerkin method

1 Introduction

In the following paper the modeling by finite element method is presented. This method is well known as very powerful for a great variety of different problems. It is also well known that a good finite element application may be developed when it exists a variational formulation for the problem which is considered. Therefore it is first of all necessary to distinguish that there are problems for which a variational formulation exists and there are also problems for which the variational formulation is not known or either a variational formulation does not exist.

Let us consider first one well known case: in elasticity problems the variational formulation associated to the system of differential equations coincides with the virtual work principle. We try, for example, to develop a rectangular element to study a case of plane stress in a thin plate with the minimum possible grade for the approximating polinomials therefore we get a non conforming element namely a formulation of an element such that the components of the deformation tensor may happen to be discontinuous on the element sides. It will be necessary to show that there are no discontinuities on the element sides to be able to accept such element as a good one.

Because the operator that we want to consider is an elliptic one we may use for this problem the same approach that may be used for a larger class of elliptic operators namely:

To determine a real function $u(x_1, x_2, ..., x_n)$ in R^n such that

$$Au = f \text{ in } \Omega(1) \tag{1}$$

$$B_j u = 0 \text{ in } \partial\Omega_j \quad (j = 1,...,r)$$

where Ω is an open bounded set of R^n, f is a real function, B_j is a linear differential operator, and $\partial\Omega_j$ ($j = 1,...,r$) are r different parts of the boundary of Ω that are supposed to be regular enough, A is a linear differential operator of order m.

If the operator A is symmetric and positive definite the problem may be shown to have a unique solution and such solution minimizes the quadratic functional

$$F(u) = (Au, u) - 2(f, u) = \tag{2}$$

$$\int_\Omega |\bar{u}(P)Au - 2\bar{u}(P)f(P)| d\Omega$$

It reverse of this theorem is also true, e.g.. the function u of the domain A that minimizes the functional (2) is the solution of the problem (1).

Existence and uniqueness of the solution are warranted in the more restricted hypothesis, that the operator A is coercive

$$(Au, u) \geq \alpha^2 \|u\|^2$$

where α is a constant $\alpha > 0$.

In a thin plate the plane xy is the average-plane and is supposed to be under distributed load $p_z(x,y)$ orthonormal to such plane and it is subject to two systems of distributed loads $p_x(x,y)$ and $p_y(x,y)$ in the average-plane.

u_x u_y u_z are the components of the displacement of the points that belong to the average plane. If u_z is small enough with respect to the thickness of the plate, the bending behaviour of the plate it is not influenced by the systems of stresses that act into the average the plane.

Let the displacements of u_z be small enough, let the material to be homogeneous and with the Kirchoff-Love hypothesis on the rotation of the average plane one obtains the well known equation for the plate bending [6]:

$$D\nabla^4 u_z = p_z$$

A₁

$$D = Et^3 / \left[12(1 - v^2)\right] \qquad \text{where}$$

$$E = \text{Young modules of matherial}$$
$$v = \text{Poisson ratio}$$

With the same hypothesis of small displacements u_x and u_y, the strain into the average plane is described by the following system of partial differential equations:

$$A_2) \quad \begin{cases} -\mu\nabla^2 u_x - \mu\gamma\dfrac{\partial}{\partial x}\left(\dfrac{\partial u_x}{\partial x} + \dfrac{\partial u_y}{\partial y}\right) = P_x \\[4mm] -\mu\nabla^2 u_y - \mu\gamma\dfrac{\partial}{\partial y}\left(\dfrac{\partial u_x}{\partial x} + \dfrac{\partial u_y}{\partial y}\right) = P_y \end{cases}$$

where $\mu = E/[2(1+v)]$

$\gamma = (1+v)/(1-v)$

In a compact form we have

$$\left[\begin{array}{c|c} A_1 & \\ \hline & A_2 \end{array}\right]\left[\begin{array}{c} u_z \\ \hat{u} \end{array}\right] = \left[\begin{array}{c} P_z \\ \hat{p} \end{array}\right] \tag{3}$$

$$\hat{u} = \begin{pmatrix} u_x \\ u_y \end{pmatrix} \qquad \hat{p} = \begin{pmatrix} P_x \\ P_y \end{pmatrix}$$

Two kinds of boundary conditions can be considered

i) $u_z = 0 \quad \dfrac{\partial u_z}{\partial n} = 0 \quad$ on $\partial\Omega$

$u_x = u_y = 0 \quad$ on $\partial\Omega$

ii) $u_z = 0 \quad \Delta^2 u_z - \dfrac{1-v}{\rho}\dfrac{\partial u_z}{\partial n} = 0 \quad$ on $\partial\Omega$

$u_x = \bar{u}_x \quad u_y = \bar{u}_y \quad$ on $\partial\Omega$

ρ is the curvature ray of $\partial\Omega$

$\bar{u}_x$ and $\bar{u}_y$ are regular functions given on the boundary $\partial\Omega$.

To be able to show existence and uniqueness of the solution of the variational problem associated to the differential problem is sufficient to prove the coerciveness of the differential operator appearing in (3).

A variational formulation is obtained by multiplying in a scalar product the two members of (3) times any vector of the test functions $v^T = (v_x, v_y, v_z)$ where $\hat{v}^T = (v_x, v_y)$ that belongs to the domain D_0 of the differential operator (3) that therefore satisfies the boundary conditions i) or ii) and this variational form is

$$(A_1 u_z, v_z) = (p_z, v_z) \quad \forall v_z \in D_0$$

$$(A_2 \hat{u}, \hat{v}) = (\hat{p}, \hat{v})$$

The two equations obtained are independent and therefore they are solved separately.

If now the discretization of the equations is considered using for simplicity the virtual work principle

$$\int_\Omega \varepsilon^T \chi \tilde{e} \, d\Omega = \int_\Omega F^T \tilde{u} \, d\Omega$$

$\tilde{\varepsilon}$ virtual deformation

$\tilde{u}$ virtual displacement

χ elasticity matrix

F   distributed loads

The term depending from the thickness is included into the matrix χ.

The domain beeing subdivided into n finite elements Ω_q $(q = 1, \dots, n)$.

The approximating relations are used

$$\underline{u} = A_q \underline{u}_q$$

$$\underline{\varepsilon} = B_q \underline{u}_q$$

$$\sum_q \left(u_q^T K_{qq} - p_q^T \right) \overline{u}_q = 0$$

$$K_{qq} = \int_{\Omega_q} B_q^T \chi B_q \, d\Omega_q$$

$$p_q^T = \int_{\Omega_q} F^T A_q \, d\Omega_q$$

$$\left(u^T K - P^T \right) \tilde{u} = 0$$

$$Ku = P$$

We note that K is the stiffness matrix and it is a symmetric matrix.

2 Example of plane stress

The case considered of plane stress is here presented by a rectangular element for a thin plate and an nondimensional coordinate system is introduced as $\xi = \dfrac{x}{a}$ and

$\eta = \dfrac{y}{b}$, where the rectangular element considered has sides a and b and these quantities may be read as initial data (see fig. 1). [1]

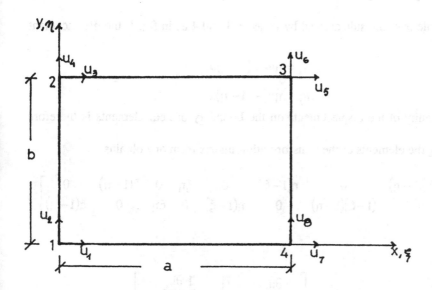

Fig. 1 The element for plane stress.

It is assumed that the displacements components have the following bilinear distribution

$$u_x = c_1\xi + c_2\xi\eta + c_3\eta + c_4 \qquad (4)$$
$$u_y = c_5\xi + c_6\xi\eta + c_7\eta + c_8$$

The constant c_1 to c_8 may be determined imposing the following conditions over the element nodes

$$u_x = u_1 \quad u_y = u_2 \quad in\ 1 \equiv (0,0)$$

$$u_x = u_3 \quad u_y = u_4 \quad in\ 2 \equiv (0,1) \qquad (5)$$

$$u_x = u_5 \quad u_y = u_6 \quad in\ 3 \equiv (1,1)$$

$$u_x = u_7 \quad u_y = u_8 \quad in\ 4 \equiv (1,0)$$

Substituting 5) into 4) inverting the expressions and reordering one obtains

$$u_x = (1-\xi)(1-\eta)u_1 + (1-\xi)\eta u_3 + \xi\eta u_5 + \xi(1-\eta)u_7 \qquad (6)$$

$$u_y = (1-\xi)(1-\eta)u_2 + (1-\xi)\eta u_4 + \xi\eta u_6 + \xi(1-\eta)u_8$$

One may notice that the developments u_x and u_y along the plate sides are linear and that they depend only on the boundary conditions over the two vertex that define the side.

For example over the side defined by verteces 3 and 4 as in fig. 1, the displacement is given by

$$u_x = \eta u_5 + (1-\eta)u_7$$

$$u_y = \eta u_6 + (1-\eta)u_8$$

The continuity of the displacements on the boundary of near elements is therefore assured.

Collecting the elements of the transformation matrix form one obtains

$$A = \begin{bmatrix} (1-\xi)(1-\eta) & 0 & \eta(1-\xi) & 0 & \xi\eta & 0 & \xi(1-\eta) & 0 \\ 0 & (1-\xi)(1-\eta) & 0 & \eta(1-\xi) & 0 & \xi\eta & 0 & \xi(1-\eta) \end{bmatrix}$$

and because
$$\varepsilon = \begin{bmatrix} \varepsilon_{xx} \\ \varepsilon_{yy} \\ \varepsilon_{xy} \end{bmatrix} = \begin{bmatrix} \dfrac{\partial u_x}{\partial x} \\ \dfrac{\partial u_y}{\partial y} \\ \dfrac{\partial u_y}{\partial x} + \dfrac{\partial u_x}{\partial y} \end{bmatrix} = \begin{bmatrix} \dfrac{1}{a}\dfrac{\partial u_x}{\partial \xi} \\ \dfrac{1}{b}\dfrac{\partial u_y}{\partial \eta} \\ \dfrac{1}{b}\dfrac{\partial u_x}{\partial \eta} + \dfrac{1}{a}\dfrac{\partial u_y}{\partial \xi} \end{bmatrix} = \hat{D}\hat{u}$$

$$\hat{D} = \begin{bmatrix} \dfrac{1}{u}\dfrac{\partial}{\partial \xi} & 0 \\ 0 & \dfrac{1}{b}\dfrac{\partial}{\partial \eta} \\ \dfrac{1}{b}\dfrac{\partial}{\partial \eta} & \dfrac{1}{a}\dfrac{\partial}{\partial \xi} \end{bmatrix} \quad \text{and} \quad \hat{u} = \begin{bmatrix} u_x \\ u_y \end{bmatrix}$$

$$u^T = \begin{bmatrix} u_1 & u_2 & u_3 & u_4 & u_5 & u_6 & u_7 & u_8 \end{bmatrix}$$

$$\varepsilon = \hat{D}Au = Bu$$

The element contribution to the stiffness matrix

$$K_\rho = \int_{\Omega_\rho} B^T \chi B d\Omega_\rho = \int_0^1 \int_0^1 \int_{-t/2}^{t/2} B^T \chi B d\xi d\eta dz$$

$$\chi = \frac{E}{1-v^2} \begin{bmatrix} 1 & v & 0 \\ v & 1 & 0 \\ 0 & 0 & \frac{1-v}{2} \end{bmatrix}$$

Operating the integration formally through Mathematica one gets the exact formulation of the element contribution to the stiffness matrix.

3 The weak formulation

There are problems for which the variational formulation does not exist.

There are problems for which the operator A may be written but it does not have such nice properties, still we may use the finite element method using a weak variational formulation that looks formally like 1) but has a quite difference significance. Still there are a great number of advantages in using the finite element method in comparison, for example, with some finite difference approach to the same problem.

The integral formulation gives advantages with respect to a formulation in terms of derivatives, there are singularities that may be overcome.

The order of derivatives used is lower with the application of Gauss divergence theorems. Some boundary conditions are included exactly into the formulation with no effort.

The boundary condition may be applied exactly on a complicated boundary. The same order of advantages is obtained when the element is developed using numerical integration instead of analytic derivation as it was done here. Therefore we may always assume a very easy generation of the stiffness matrix.

All these reasons take us to try a generalised formulation as given in the form

$$(Au,v) = (f,v) \tag{7}$$

also for problems that do not present such nice properties as the problem above.

We may assume to write

$$\int_\Omega v^T Au d\Omega + \sum_j \int_{\partial\Omega_j} v^T B_j u d\partial\Omega_j = \int_\Omega v^T f d\Omega \tag{8}$$

In many cases it is possible to apply the Gauss divergence theorems and obtain a *weak formulation* where it is not proved the existence of the minimum as in the strong formulation but is still possible to say something on the problem.

If the integrals into the general formula above are integrals obtained by the method of weighted residuals one may see that many advantages seen above may still be obtained.

Taking into evidence the weights used one obtains

$$\int_{\Omega} \omega R d\Omega = 0$$

where ω is the weight function and R the residual due to the substitution of the exact solution by an approximate solution.

Many different methods may be used

- Collocation Method
- Least Squares Method
- Moment Method (mostly for one dimensional problems, because in such field the orthogonal polynomial theory is well known)
- Galerkin Method   in such method the weight funtions ω_j are chosen as the shape functions obtained by the approximation applied such that one obtains

$$\int_{\Omega} N_j(x) R d\Omega = 0 \qquad j = 1,...,m$$

Using the well known fact that continuous functions vanish if they are orthogonal with respect to each member of a complete set, it is possible to show that the Galerkin method makes the residual to vanish by building it orthogonal to each member of a complete set of basis functions as $[N_j]$.

In the following we use this last method for two very different applications.

4 The analysis of a tidal flow in a lagoon

The Navier-Stokes equations for a viscous incompressible flow can be considerably simplified by introducing a number of basic assumptions such as the "long wave" modelling which allows neglecting of vertical velocity and acceleration thus reducing the problem to a two-dimensional one, or the absence of viscosity as in the present case.

The hypothesis are that
  a) the vertical effects (acceleration and diffusion) are negligible
  b) horizontal velocities do not vary with the depth
  c) the water elevation with respect to the free surface at rest is at any time
     much smaller than the depth
  d) convective and viscous terms are negligible.

After integration over the depth according to assumptions a) and b) and the application of Gauss divergence theorem, the integral form of the governing equations results to be given by

$$\frac{\partial}{\partial t}\left(\int_{\Omega} U d\Omega\right) = -\int_{\Gamma}(F_1 n_1 + F_2 n_2)d\Gamma + \int_{\Omega} R_s d\Omega$$

where Ω is the two dimensional domain of interest, $\Gamma = \partial\Omega$ is its boundary with outer unit normal $n = [n_1 n_2]^T$ and

$$U = [\eta \quad u_1 \quad u_2]^T \qquad F_1 = [Hu_1 \quad g\eta \quad 0]^T \qquad F_2 = [Hu_2 \quad 0 \quad g\eta]^T$$

$$R_s = \left[0 \quad fu_2 - \frac{g|u|u_1}{C^2 H} \quad -fu_1 - \frac{g|u|u_2}{C^2 H}\right]$$

here η is the water elevation, u_1 and u_2 are the velocities in the x and y directions, H is the water depth and g, f and C respectively, the gravity acceleration, the Coriolis and the Chezy coefficients.

The set of equations to be solved is therefore

$$\frac{\partial U}{\partial t} + \frac{\partial F_1}{\partial x_1} + \frac{\partial F_2}{\partial x_2} = R_s$$

or in extended form
$$\begin{cases} \dfrac{\partial \eta}{\partial x} + \dfrac{\partial(Hu_1)}{\partial x_1} + \dfrac{\partial(Hu_2)}{\partial x_2} = 0 \\[2mm] \dfrac{\partial u_1}{\partial t} + g\dfrac{\partial \eta}{\partial x_1} = gu_2 - \dfrac{g|u|u_1}{C^2 H} \\[2mm] \dfrac{\partial u_2}{\partial t} + g\dfrac{\partial \eta}{\partial x_2} = gu_1 - \dfrac{g|u|u_2}{C^2 H} \end{cases}$$

A semi - implicit method is used to solve such system of equations

$$F_i = F_i^* + F_i^{**}$$
$$U^{(n+1)} = U^{(n)} + \Delta U^* + \Delta U^{**}$$

ΔU^* and ΔU^{**} are the corresponding increments of the solution vector and $U^{(n)}$ and $U^{(n+1)}$ denote the global solution vectors at the n-th and (n+1)-th time respectively and $F_i^* = 0$ and $F_i^{**} = F_i$ initially.

Therefore the equations are split into

$$\frac{\partial \Delta U^*}{\partial t} + \frac{\partial F_1^*}{\partial x_1} + \frac{\partial F_2^*}{\partial x_2} = R_S \tag{9}$$

$$\frac{\partial \Delta U^{**}}{\partial t} + \frac{\partial F_1^{**}}{\partial x_1} + \frac{\partial F_2^{**}}{\partial x_2} = 0 \tag{10}$$

which are integrated in time, in unit, by using an explicit Taylor-Galerkin method for (9) and an implicit ϑ method for (10). Equations (9) are first discretized in time via the second order Taylor expansion for a time step Δt

$$\left(\Delta U^*\right)^{n+1} = \Delta t \left(\frac{\partial \Delta U^*}{\partial t}\right)^{(n)} + \frac{\Delta t^2}{2}\left(\frac{\partial^2 \Delta U^*}{\partial t^2}\right)^{(n)}$$

which can be rewritten since

$$\frac{\partial \Delta U^*}{\partial t} = R_S$$

and

as

$$\left(\Delta \qquad\qquad\qquad\qquad\qquad\qquad)^n \tag{11}$$

where

$$G = \frac{\partial R_S}{\partial \Delta U^*}$$

Due to the computational complexity of evaluating the right hand side of (11) the two-step version of the Taylor-Galerkin algorithm is here adopted.

This consist of:

approximating the value of $U^{(n+\frac{1}{2})}$ by the linear Taylor expansion

$$U^{(n+\frac{1}{2})} = U^{(n)} + \frac{\Delta t}{2}\left(R_S\right)^{(n)}$$

from which $R_S^{(n+\frac{1}{2})}$ can be evaluated while the same expansion for $R_S^{(n+\frac{1}{2})}$ would give

$$R_S^{(n+\frac{1}{2})} = R_S^{(n)} + \frac{\Delta t}{2}\left(\frac{\partial R_S}{\partial t}\right)^{(n)} = (R_S)^{(n)} + \frac{\Delta t}{2}(GR_S)^{(n)}$$

when $(GR_S)^{(n)}$ is obtained as

$$(GR_S)^{(n)} = \frac{2}{\Delta t}\left(R_S^{(n+\frac{1}{2})} - R_S^{(n)}\right).$$

The finite element approximation in space of the above quantities is performed by using linear weighting (shape) functions at full time steps $(n, n+1, ...)$ and constant functions at half time steps $(n-1/2, n+1/2, ...)$. Hence the time stepping process takes for each node j the form

$$\left(M\Delta U^*\right)_j^{(n+1)} = \Delta t \int_\Omega (R_S)^{(n)} \varphi_j d\Omega + \frac{\Delta t^2}{2}\int_\Omega \frac{2}{\Delta t}\left[(R_S)^{(n+\frac{1}{2})} - (\overline{R}_S)^{(n)}\right]\varphi_j d\Omega =$$

$$= \Delta t \int_\Omega (R_S)^{(n+\frac{1}{2})} + \left[(R_S)^{(n)} - (\overline{R}_S)^{(n)}\right]\varphi_j d\Omega \qquad j = 1,...,N$$

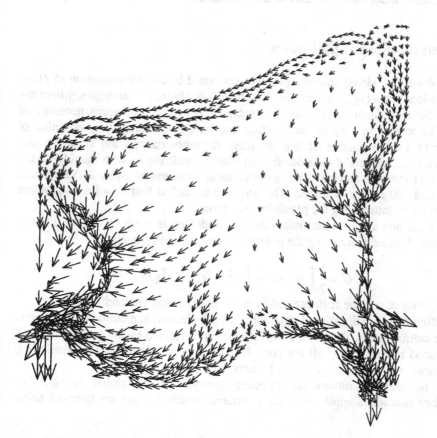

Fig. 2 The flow in the central Venice lagoon at 6 hours simulations.

where N is the total numbers of nodes in the mesh, φ_j is the linear weighting function at that node, the bar denotes averages over the element and

$$M = [M]_{ij} = \left[\int_\Omega \varphi_i \varphi_j d\Omega \right]_{ij} \qquad i, j = 1, ..., N$$

is the mass matrix.

Equation solving and core optimization is performed over such system of equations. The Solution is obtained by the conjugate gradient method which has resulted quite efficient in all applications.

A first example of application has been the central part of Venice lagoon. The mesh has 1581 elements and 890 nodes. The water depth varies from 0.8 metres in the inner region to about 14.0 metres in the region of finer mesh subdivision and a time step of two minutes is used. The complete Venice lagoon is analysed comprising 1967 nodes and 3423 elements, a time step of 30 secs is required in this case for stability. The conjugate gradient method shows to be very well performing even in time stepping when the Cholesky decomposition only requires a forward and a backward substitution. In fig. 2 the central lagoon is shown with the appearence of motion after 6 hours of motion [2,3].

4 An elasto-plastic application

The problem considered now is a problem generated by the determination of plane stress induced in a big cavity created into the Cava Madre in Candoglia, where the marble for the Duomo of Milano is extracted. There is a continuous necessity of extraction to remake parts of the building that have been damaged by pollution of the air or by the injuries of the weather. A transversal section is taken into consideration. The model considered is an elasto-plastic model with bounds for the resistence to traction and cut. The cavity is about 40 metres of eight, 21.5 metres of width, and 100 metres of depth. The problem is studied first as a linear problem and second by introducing the plasticity conditions.

The condition to create the finite element modelling is given by the equilibrium equations obtained minimising the potential energy of the system e.g.

$$U = \frac{1}{2} \int_V \sigma^T \varepsilon dV - \int_V u^T w dV - \int_S u^T q dS$$

that are respectively the stiffness matrix of the element and the loading vector.

Starting from the discretized equations of each element one builds the equations for the entire system. The procedure, for what is concerned by the global matrix K is obtained by assembling all the contribution of each element in the positions ij determined with respect to the global numbering.

The boundary conditions are imposed acting over the global matrix. The boundary conditions applied are the geometric conditions and are imposed using

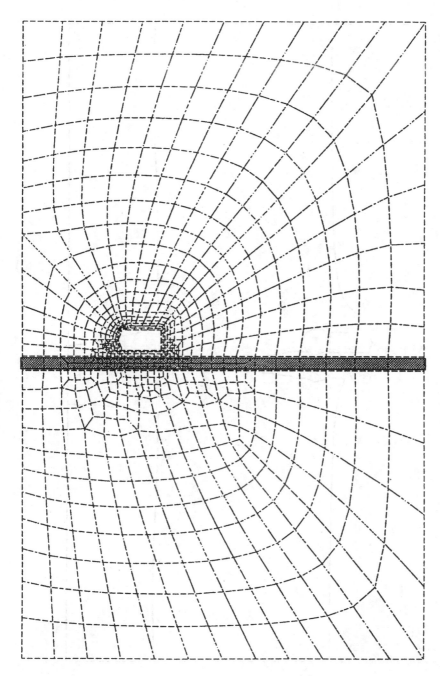

Fig. 3 The fracture of the quarry.

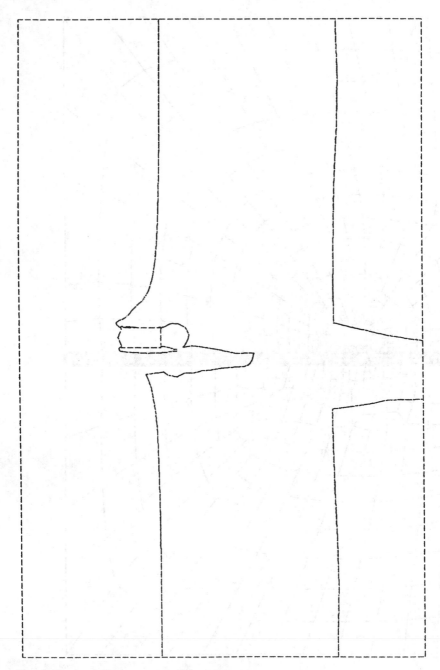

Fig. 4 The second component of the stress.

prescribed values for the displacement of a node. If, as very often happens, the imposed value is zero for a constraint reaction one will eliminate the row or the column corresponding to that variable. The linear system generated has been solved by a conjugate gradient procedure. For the plasticity the equtions considered are the usual referring to three invariants of the tension J_1, J_2, J_3 [4,5], where ε is the vector of the strain, ρ is the vector of the stress, w is the vector of the volume forces and q is the vector of the surface forces, V is the region of the application of the model and S is its boundary.

$$\frac{\partial U^\rho}{\partial u^\rho} = 0$$

$$\int_{V^\rho} \left(B^T CB\right)u^\rho dV - \int_{V^\rho} N^T w dV - \int_{S^\rho} N^T q dS = 0$$

from which one obtains the system to be solved

$$K^\rho u^\rho - F^\rho = 0$$

$$K^\rho = \int_{V^\rho} B^T CB dV$$

$$F^\rho = \int_{V^\rho} N^T w dV + \int_{V^\rho} N^T q dS .$$

The results are indicated in Fig. 3 where the zone considered is presented and in Fig. 4 where the second components of the stress is plotted.

Acknowledgements

This work has been done with the support of the C.N.R., within "Progetto Strategico Sistema Lagunare Veneziano".

References

1. Morandi Cecchi, M., Lami, C.: Studio e sviluppo di un elemento finito rettangolare piano per lamine tensoinflesse. Rendiconti Ist. Lombardo Acc. Sc. Lett. 106, (1972)
2. Morandi Cecchi, M.: Study of the behaviour of oscillatory waves in a lagoon. Int. J. Num. Meths. Engng 27, 103-112 (1989)
3. Morandi Cecchi, M., Pica, A.: General circulation modelling in shallow water problems with data assimilation computer techniques and applications. In: W.R. Blain, Cabrera E. (eds.) Hydraulic Engineering Software IV. Proceedings. Comput. Mechs. 249-262. New York: Elsevier 1992
4. Prager W., Hodge, P.G.: Theory of perfectly plastic solids. New York: Wiley 1951
5. Drucker, D.C., Prager, W.: Soil mechanics and plastic analysis on limit design. Q. Appl. Math., 10, 157-165 (1992)
6. Fung, Y.C.: Foundations of Solid Mechanics. New York: Prentice Hall 1965

References

Finite Element Method by Using Mathematica

M. D. Mikhailov

Institute for Applied Mathematics and Informatics, Sofia, Bulgaria

Abstract. The author is developing a series of *Mathematica* notebooks intended to serve a large variety of students, designers, engineers, and managers who need to use finite element analysis. The notebooks emphasise how to apply simple but universal programs to a wide variety of engineering disciplines. The multidisciplinary nature of finite element analysis is showed by applications for thermal, fluid, electrical, and structural engineers. Steady state, periodic, spectral, and transient problems are included in the notebooks. This presentation contains a mixture of selected materials from several notebooks. Only one-dimensional steady state problems are demonstrated.

Keywords. Mathematica, finite elements, steady state method, field problems, beam deflection, composite medium, radiation shields

1 Introduction

The finite element method is a widely accepted engineering tool that has many applications. For this reason textbooks on the subject abound and most engineering schools offer courses on the finite element method. The teaching of the finite element method is a necessity in curricula for thermal, fluid, electrical, and structural engineers.

Most of the finite element textbooks currently in use [1-3] emphasise the basic theoretical aspects of the method with applications being presented without explaining the programming aspects of the technique. The textbooks devoted to finite element programming [4-6] contain many unimportant details because of the FORTRAN language that is used.

In the last years a new generation of software has appeared that will completely change the education process. Such packages are called computer algebra systems (CAS) in general. Among these software packages, *Mathematica* and *Maple* are well known over the world since they allow the creation of notebooks mixing text, animated graphics, and actual input. The CAS, *Mathematica,* handles numerical, symbolic, and graphical computations in a unified way. Detailed information is given in [7].

The author is developing a series of Mathematica notebooks intended to serve a large variety of students, designers, engineers, and managers who need to use finite element analysis. The notebooks enable the user to understand deeply the method and to obtain practical experience of finite element computations.

The notebooks emphasize how to apply simple but universal programs to a wide variety of engineering disciplines. The multidisciplinary nature of finite element analysis is showed by applications for thermal, fluid, electrical, and structural engineers. Many steady state, periodic, spectral, and transient problems are presented to illustrate how the finite element method is used as a practical tool.

The programs are simple, easy to explain and universal. For a new type of physical problem only local vectors and local matrices have to be specified. The programs are designed for teaching purposes. More efficient versions for engineering computations are under development.

The aim of this presentation is to describe briefly a new approach of teaching the finite element method by using the *Mathematica* package. This notebook is a mixture of selected materials from several notebooks. Only one-dimensional steady state problems are demonstrated.

2 Steady State Finite Element Method

Calculate the temperature distribution through the wall shown below, where k1 and k2 are thermal conductivity, h is the convection coefficient, T_f is the ambient temperature and T_s is the surface temperature [1].

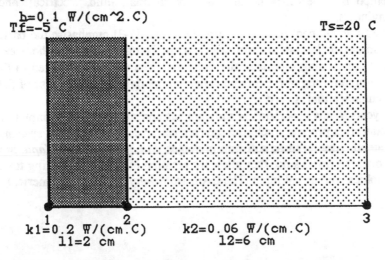

The *Finite Element model* consists of a collection of domain elements and boundary elements interconnected at a series of nodes. The elements are pieces of various sizes and shapes. They could have different numbers of nodes. When the domain elements are volumes, surface, or lines, the corresponding boundary elements are surfaces, lines or nodes. For identification, nodes are marked by numbers.

Domain elements are specified by the list domain. The above problem has two elements {1,2} and {2,3}.

In[1]:=
domain={{1,2},{2,3}};

The **boundary elements** are nodes with unknown temperature specified by the list boundary.

In[2]:=
boundary={{1}};

The points with **known** temperatures are specified by the list known.

In[3]:=
known={3};

The known temperatures are specified by the list **value**.

In[4]:=
value={Ts};Ts=20;

One or more variables are associated with each node. The number of variables per node is called **degree of freedom** (dof). In thermal problems, we have one degree of freedom per node (temperature). In structural problems, we have up to 6 degrees of freedom per node (three translations and three rotations).

In[5]:=
dof=1;

The **variable indices** are defined by the rule:

In[6]:=
indices[nodeNumber_] :=
**   Table[dof*nodeNumber-(dof-i),{i,dof}]**

The **variable topology** of the model is defined by the rule:

In[7]:=

```
variablesTopology:=Table[
  Flatten[Table[indices[domain[[e,i]]],
          {i,Length[domain[[e]]]}]],
                {e,Length[domain]}]
```

In[8]:=

```
variablesTopology
```

Out[8]=

```
{{1, 2}, {2, 3}}
```

The number of *all variables* is

In[9]:=

```
allVariables:=Last[Union[
  Flatten[variablesTopology]]]
```

In[10]:=

```
allVariables
```

Out[10]=

3

The domain local matrix of an element with n nodes has $n \times dof$ by $n \times dof$ components. The local vector of an element with n nodes has $n \times dof$ components.

The *domain local vector* and *domain local matrix* for the examples considered are

In[11]:=

```
localVector["fd",e_]:={0,0}
localMatrix["Kd",e_]:=
      (k/l)[[e]]{{1,-1},{-1,1}}
```

where

In[13]:=

```
k={0.2,0.06}; l={2,6};
```

In[14]:=

```
localMatrix["Kd",1]
```

Out[14]=

```
{{0.1, -0.1}, {-0.1, 0.1}}
```

In[15]:=
localMatrix["Kd",2]

Out[15]=
{{0.01, -0.01}, {-0.01, 0.01}}

The *boundary local vector* and *boundary local matrix* for all the examples in this notebook are

In[16]:=
localVector["fb",e_]:=fi[[e]]{1}
localMatrix["Kb",e_]:=al[[e]]{{1}}

where

In[18]:=
fi={h Tf};   h=0.1; Tf=-5; al={h};

In[19]:=
localMatrix["Kb",1]

Out[19]=
{{0.1}}

The matrix `localGlobal` transforms a local vector to an extended local vector. The number of rows is equal to the number of element variables. The number of columns is equal to the number of all variables.

In[20]:=
```
localGlobal[ region_,e_]:=
   Table[ If[ j==region[[e,i]],1,0],
           {i,Length[ region[[e]]]},
           {j,allVariables}]
```

In[21]:=
localGlobal[variablesTopology,2]

Out[21]=
{{0, 1, 0}, {0, 0, 1}}

The *extended vector* is

In[22]:=
```
extendedVector[ name_,region_,e_]:=
   Transpose[ localGlobal[ region,e]].
             localVector[ name,e]
```

In[23]:=
```
extendedVector["fd",variablesTopology,1]
```

Out[23]=
```
{0, 0, 0}
```

In[24]:=
```
extendedVector["fd",variablesTopology,2]
```

Out[24]=
```
{0, 0, 0}
```

In[25]:=
```
extendedVector["fb",boundary,1]
```

Out[25]=
```
{-0.5, 0, 0}
```

The *extended matrix* is

In[26]:=
```
extendedMatrix[name_,region_,e_]:=
  Transpose[localGlobal[region,e]].
             localMatrix[name,e].
             localGlobal[region,e]
```

In[27]:=
```
extendedMatrix["Kd",variablesTopology,1]
```

Out[27]=
```
{{0.1, -0.1, 0}, {-0.1, 0.1, 0}, {0, 0, 0}}
```

In[28]:=
```
extendedMatrix["Kd",variablesTopology,2]
```

Out[28]=
```
{{0, 0, 0}, {0, 0.01, -0.01},
  {0, -0.01, 0.01}}
```

In[29]:=
```
extendedMatrix["Kb",boundary,1]
```

Out[29]=
```
{{0.1, 0, 0}, {0, 0, 0}, {0, 0, 0}}
```

The *global vector* is the sum of the extended local vectors

The *global vector* is the sum of the extended local vectors

In[30]:=
```
globalVector[name_,region_]:=
  Sum[extendedVector[name,region,e],
                    {e,Length[region]}]
```

In[31]:=
```
globalVector["fd",variablesTopology]
```
Out[31]=
```
{0, 0, 0}
```

In[32]:=
```
globalVector["fb",boundary]
```
Out[32]=
```
{-0.5, 0, 0}
```

The *global matrix* is the sum of the extended local matrices

In[33]:=
```
globalMatrix[name_,region_]:=
  Sum[extendedMatrix[name,region,e],
                  {e,Length[region]}]
```

In[34]:=
```
globalMatrix["Kd",variablesTopology]
```
Out[34]=
```
{{0.1, -0.1, 0}, {-0.1, 0.11, -0.01},
  {0, -0.01, 0.01}}
```

In[35]:=
```
globalMatrix["Kb",boundary]
```
Out[35]=
```
{{0.1, 0, 0}, {0, 0, 0}, {0, 0, 0}}
```

The *model vector* is the sum of the domain and boundary global vectors

In[36]:=
```
modelVector:=
   globalVector[ "fd",variablesTopology]+
   globalVector[ "fb",boundary]
```

In[37]:=
```
f=modelVector
```

Out[37]=
```
{-0.5, 0, 0}
```

The *model matrix* is the sum of the domain and boundary global matrices

In[38]:=
```
modelMatrix:=
   globalMatrix[ "Kd",variablesTopology]+
   globalMatrix[ "Kb",boundary]
```

In[39]:=
```
m=modelMatrix
```

Out[39]=
```
{{0.2, -0.1, 0}, {-0.1, 0.11, -0.01},
   {0, -0.01, 0.01}}
```

The model algebraic system is *modified* whenever some of the variables are known. The indices and the values of these variables are specified in the lists **known** and **value**.

The modification of the system $m \cdot x = f$ for a known $x(n)$ is as follows:

1. The vector f is modified by subtracting from f the column n of the matrix m multiplied by $x(n)$ and then setting $f(n) = x(n)$.

2. All of the coefficients in row and column n are set equal to zero except the diagonal term, which is set to one.

In[40]:=
```
modify:=
  Do[ changeSystem[known[[i]],value[[i]]].
                    {i,1,Length[known]}]
changeSystem[n_,x_]:=
  (f=f-Transpose[m][[n]]*x;
     f[[n]]=x;
     Table[m[[n,j]]=0,{j,Length[m]}];
     Table[m[[i,n]]=0,{i,Length[m]}];
     m[[n,n]]=1)
```

In[42]:=
```
modify
```
In[43]:=
```
MatrixForm[m]
```

Out[43]://MatrixForm=

0.2	-0.1	0
-0.1	0.11	0
0	0	1

In[44]:=
```
f
```

Out[44]=
{-0.5, 0.2, 20}

The rule steady finds the solution by solving the algebraic system obtained

In[45]:=
```
steady:=(m=modelMatrix;f=modelVector;
          modify; LinearSolve[m,f])
```

The finite element method give the temperature at the nodal points

In[46]:=
```
solution=steady
```

Out[46]=
{-2.91667, -0.833333, 20.}

The finite element method assumes that the temperature distribution in a two point line element is linear. The femPlot1D plots this distribution by using data specified in the lists l and solution.

In[47]:=
```
femPlot1D:=Module[{coordinate,x},
  coordinate:=(x[1]=0;
    x[i_]:=x[i-1]+l[[i-1]];
    Table[x[i],{i,1,Length[l]+1}]);
 ListPlot[Table[
    {coordinate[[i]],solution[[i]]},
      {i,Length[solution]}],
      PlotJoined->True,
      AxesLabel->{"x","solution"}]]
```
In[48]:=
```
femPlot1D;
```

solution

3 One-Dimensional Field Problems

The one-dimensional field model has many important applications such as the heat flow through a fin or composite wall, the deflection of simply supported beams, etc. The model is described by

$$(kP')' - bP + g = 0$$
$$\alpha_0 P - \beta_0 P' = \varphi_0$$
$$\alpha_1 P + \beta_1 P' = \varphi_1$$

The *local element matrices* and *local element vectors* are:

In[49]:=

```
localMatrix[ "Kd",e_]:=
   (k/l)[[e]]{{1,-1},{-1,1}}+
   (b l/6)[[e]]  {{2,1},{1,2}}
localVector[ "fd",e_]:=(g l/2)[[e]]{1,1}
```

Temperature $T(x)$ in a one dimensional ***rectangular fin*** is defined by

$$kT'' - \frac{hP}{A}(T - T_f) = 0$$

$$T(0) = T_0$$

$$-kT'(L) = h(T(L) - T_f)$$

where h is the convection coefficient, P is the distance around the fin, k is the thermal conductivity, A is the cross-sectional area, T_0 is the known temperature, T_f is the air temperature.

Calculate the temperature distribution in a rectangular fin 8 cm long, 4 cm wide, and 1 cm thick.

This problem is a special case from the model:

$$b = hP/A, \ g = bT_f, \ \alpha_0 = 1$$

$$\beta_0 = 0, \ \varphi_0 = T_0, \ \alpha_1 = h, \ \beta_1 = 1, \ \varphi_1 = hT_f$$

$$k=3 \ W/(cm.C)$$

$$h=0.1 \ W/(cm^2.C)$$

$$Tf= 20 \ C$$

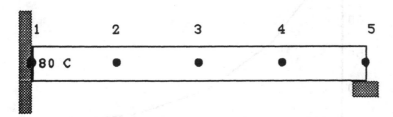

8 cm

The input data are:

In[51]:=
```
ke=3;he=0.1;T0=80;Tf=20;le=2;P=10;A=4;
l={1,1,1,1}le;
k={1,1,1,1}ke;
b={1,1,1,1}he*P/A;
g=b*Tf;
domain={{1,2},{2,3},{3,4},{4,5}};
boundary={{5}};
al={he} ;
fi={he Tf};
known={1};
value={T0};
```

The solution is

In[52]:=
```
solution=steady
femPlot1D;
```

Out[52]=
{80., 53.9456, 39.8719, 32.8119, 30.2737}

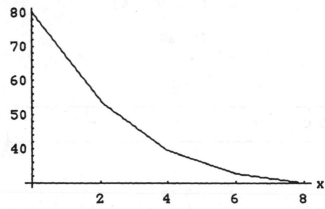

3.1 Beam Deflection

The model deflection curve is defined by:

$$E \ I \ w'' = -M \ (x)$$
$$w(0) = 0$$
$$w(L) = 0$$

where E is the elastic modules of the material, I is the area moment of the cross section, M (x) is the internal bending moment and w is the deflection.

Calculate the deflection of a beam shown below. It has been reinforced over the centre one-half of its span through the use of steel plates that are welded to the basic section [1]. The beam data and the finite element model are given in the figure.

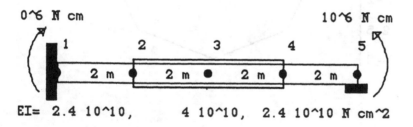

The input data are:

In[54]:=
```
l={1,1,1,1}200;
k={2.4, 4, 4, 2.4} 10^10;
b={1,1,1,1}0;
g=-{1,1,1,1}10^6;
domain={{1,2},{2,3},{3,4},{4,5}};
known={1,5};
value={0,0};
Clear[boundary,al,fi]
```

The solution is

In[72]:=
```
solution=steady
femPlot1D;
```
Out[72]=
```
{0., -2.5, -3., -2.5, 0.}
```

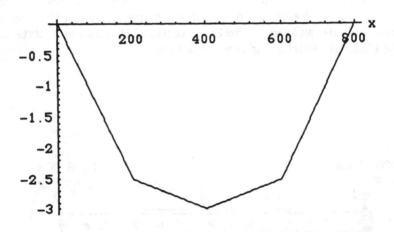

3.2 Composite Medium

The finite element method is used to calculate the interface temperatures and heat flow rate through a composite medium consisting of several layers of slabs, coaxial cylinders, and concentric spheres.

The local domain and boundary matrices and vectors are for *a slab* are

In[74]:=
```
localMatrix["Kd",e_]:=
    (k/l)[[e]]{{1,-1},{-1,1}}
localVector["fd",e_]:= {0,0}
```

where:k is the list of thermal conductivity, l is the list of lengths.

The local vector is the same for all problems of composite medium and radiation shields. The *heat flow rate through a plane wall* is

In[76]:=
```
heatFlowRate:=(k/l)[[1]]*A*
               (T[[1]]-T[[2]])
```

Wall example 1. Calculate the surface temperatures for an plate of thickness 2 cm with thermal conductivity $20\dfrac{W}{m\,C}$ subjected to a constant, uniform flux $10^5\dfrac{W}{m^2}$ at one boundary surface. From the other boundary surface, heat is dissipated by convection into a fluid at temperature 50 C with a heat transfer coefficient $500\dfrac{W}{m^2C}$ [8].

In[77]:=
```
A=1;
l={0.02};
k={20};
domain={{1,2}};
boundary={{1},{2}};
al={0,500};
fi={10^5,500*50};
Clear[known,value]
steady
```

Out[85]=
```
{350., 250.}
```

Wall example 2. The wall consists of a fire clay brick of thickness 0.2 m, thermal conductivity $1\dfrac{W}{m\,C}$ and an insulation of thickness 0.03 m, thermal conductivity $0.05\dfrac{W}{m\,C}$. The inside and outside surface of the wall are kept at 830 C and 30 C, respectively. Calculate the interface temperature (between the wall and the insulation) and the heat transfer rate across the wall per metre square [8].

In[86]:=
```
Clear[boundary,al,fi]
A=1;
l={0.2,0.03} ;
k={1,0.05} ;
domain={{1,2},{2,3}};
known={1,3};
value={830,30};
T=steady
heatFlowRate
```

Out[93]=
```
{830., 630., 30.}
```

Out[94]=
```
1000.
```

The local domain matrix for a *cylinder* is:

In[95]:=
```
localMatrix["Kd",e_]:=
  (k[[e]]/Log[r[[e+1]]/r[[e]]])*
                {{1,-1},{-1,1}}
```
where: k is the list of thermal conductivity, r is the list of radius.

The *heat flow rate* through a *cylinder* is:

In[96]:=
```
heatFlowRate:=2*N[Pi,10]*h*
  (k[[1]]/Log[r[[2]]/r[[1]]])*
                (T[[1]]-T[[2]])
```
where: h is the length of the cylinder, T[[1]] and T[[2]] correspond to the surface temperatures of the first layer.

Cylinder example. A pipe of outside radius 4 cm is covered with a

1 cm thick asbestos insulation of thermal conductivity $0.15\dfrac{W}{m\ C}$,

which is in turn covered with a 3 cm cm thick fibre glass insulation of

thermal conductivity $0.05\dfrac{W}{m\ C}$. The pipe is kept at 330C and the outside surface of the insulation is at $30\,C$. Determine the heat transfer rate per meter length and the interface temperature between the asbestos and the fibre glass insulation [8].

In[97]:=

```
Clear[boundary,al,fi]
h=1;
r={0.04,0.05,0.08} ;
k={0.15,0.05} ;
domain={{1,2},{2,3}};
known={1,3};
value={330,30};
T=steady
heatFlowRate
```

Out[104]=

{330., 289.01, 30.}

Out[105]=

173.127

The local domain matrix for a *sphere* is:

In[106]:=

```
localMatrix["Kd",e_]:=
  (k[[e]]/(1/r[[e]]-1/r[[e+1]]))*
                  {{1,-1},{-1,1}}
```

The *heat flow rate* through a *sphere* is:

In[107]:=

```
heatFlowRate:=4*N[Pi,10]*
   (k[[1]]/(1/r[[1]]-1/r[[2]]))*
                  (T[[1]]-T[[2]])
```

where T[[1]] and T[[2]] correspond to the surface temperatures of the first layer.

Sphere example. A hollow nickel-steel sphere has an inner radius $10\,cm$, thickness $5\,cm$ and thermal conductivity $12\dfrac{W}{m\,C}$. It is covered with a $2\,cm$ thick insulation of thermal conductivity $0.5\dfrac{W}{m\,C}$. The inner surface of the sphere is kept at $230C$ and the other surface of the insulation is dissipating heat by convection into a surrounding at $30C$ with a heat transfer coefficient $20\dfrac{W}{m^2C}$. Calculate the total heat flow rate across the sphere and the interface temperature [8].

In[108]:=
```
r={0.1,0.15,0.17} ;
k={12,0.5} ;
domain={{1,2},{2,3}};
boundary={{3}};
al={20*0.17^2};
fi={20*30*0.17^2};
known={1};
value={330};
T=steady
heatFlowRate
```

Out[116]=
```
{330., 306.7, 175.122}
```

Out[117]=
```
1054.08
```

3.3 Radiation Shields

Finite element method is used to examine the radiation heat transfer between two parallel plates, long coaxial cylinders and concentric spheres and the resulting reduction in the heat transfer rate when one or more layers of radiation shields are placed between them.

The problem is reduced to the network system which potentials have values

In[118]:=
```
value:=(temperatures/100)^4
```

Once the solution for potentials is found the temperatures are

In[119]:=
```
shieldsTemperatures:=
    (T=steady;Sqrt[Sqrt[T]]*100)
```

The local domain matrix for *parallel plate* is

In[120]:=
```
localMatrix["Kd",e_]:=
    A/(1/emissivity[[e,1]]+
      1/emissivity[[e,2]]-1)*
              {{1,-1},{-1,1}}
```

where : A is an area, emissivity is the list of the emissivities of the inner and outer surfaces. The heat flow rate through plane wall is

In[121]:=
```
heatFlowRate:=5.67*A*(T[[1]]-
    T[[2]])/(1/emissivity[[1,1]]+
            1/emissivity[[1,2]]-1)
```

where T[[1]] and T[[2]] correspond to the temperatures of the first and second shields.

Plane Shield example 1. Calculate the radiation heat transfer rate between two large parallel plates. The first plate is at 800C and has an emissivity 0.9. The second plate is at 300C and has an emissivity 0.5 [8].

```
In[122]:=
Clear[boundary,al,fi]
A=1;
emissivity={{0.9,0.5}} ;
domain={{1,2}};
known={1,2};
temperatures={800,300};
shieldsTemperatures
heatFlowRate
```

Out[128]=

```
{800, 300}
```

Out[129]=

```
10783.4
```

Plane Shield example 2. Consider ***Plane Shield example 1*** when a
radiation shield of emissivity 0.1 at both surfaces is placed between the
plates [8]

```
In[130]:=
emissivity={{0.9,0.1},{0.1,0.5}} ;
domain={{1,2},{2,3}};
known={1,3};
temperatures={800,300};
shieldsTemperatures
heatFlowRate
```

Out[134]=

```
{800., 682.757, 300.}
```

Out[135]=

```
1078.34
```

The local domain matrix for a ***long cylinder*** is

```
In[136]:=
localMatrix["Kd",e_]:= N[Pi,10]*L/
    (1/emissivity[[e,1]]/diameter[[e]]+
    (1/emissivity[[e,2]]
                    -1)/diameter[[e+1]])*
    {{1,-1},{-1,1}}
```

where : L is a length, **emissivity** is the list of the emissivities of the inner and outer surfaces, **diameter** is the list of the diameters. The heat flow rate for a long cylinder is

In[137]:=
```
heatFlowRate:=5.67*N[Pi,10]*L*
   (T[[1]]-T[[2]])/
   (1/emissivity[[1,1]]/diameter[[1]]+
   (1/emissivity[[1,2]]-1)/diameter[[2]])
```

where **T[[1]]** and **T[[2]]** correspond to the temperatures of the first and second shields.

Cylindrical Shield example 1. Calculate the radiation heat flow rate per metre length between two long coaxial cylinders. The inner cylinder has a diameter 5 cm, emissivity 0.9 and temperature 1000 K. The outer cylinder has a diameter 10 cm, emissivity 0.8 and temperature 500 K [8].

In[138]:=
```
L=1;
diameter={0.05,0.1};
emissivity={{0.9,0.8}} ;
domain={{1,2}};
known={1,2};
temperatures={1000,500};
shieldsTemperatures
heatFlowRate
```

Out[144]=
{1000, 500}

Out[145]=
6754.87

Cylindrical Shield example 2. Solve the Example 2 when a shield is placed between the two cylinders. The shield has a diameter 7 cm, emissivity 0.05 for the left surface and emissivity 0.1 for the right surface.

In[145]:=
```
diameter={0.05,0.07,0.1};
emissivity={{0.9,0.05},{0.1,0.8}} ;
domain={{1,2},{2,3}};
known={1,3};
temperatures={1000,500};
shieldsTemperatures
heatFlowRate
```

Out[151]=
```
{1000., 781.45, 500.}
```

Out[152]=
```
380.392
```

The element matrices for a *concentric spheres* are

In[153]:=
```
localMatrix["Kd",e_]:= N[Pi,10]/
   (1/emissivity[[e,1]]/diameter[[e]]^2+
   (1/emissivity[[e,2]]-
               1)/diameter[[e+1]]^2)*
               {{1,-1},{-1,1}}
```

where : `emissivity` is the list of the emissivities of the inner and outer surfaces, `diameter` is the list of the diameters. The heat flow rate through concentric spheres is

In[154]:=
```
heatFlowRate:=5.67*N[Pi,10]*
   (T[[1]]-T[[2]])/
   (1/emissivity[[1,1]]/diameter[[1]]^2+
   (1/emissivity[[1,2]]
               -1)/diameter[[2]]^2)  w
```

here `T[[1]]` and `T[[2]]` correspond to the temperatures of the first and second shields.

Spherical Shield example. Calculate the radiation heat flow rate between two coaxial sphere. The inner sphere has a diameter 5 cm, emissivity 0.9 and temperature 800 K. The outer sphere has a diameter 10 cm, emissivity 0.8 and temperature 300 K. Two shields are placed between the sphere. One of the shields has a diameter 7 cm, emissivity

In[155]:=

```
diameter={0.05,0.07,0.08,0.1};
emissivity={{0.9,0.05},
            {0.1,0.06},{0.07,0.8}};
domain={{1,2},{2,3},{3,4}};
known={1,4};
temperatures={800,300};
shieldsTemperatures
heatFlowRate
```

Out[160]=

{800., 709.089, 547.693, 300.}

Out[161]=

6.46181

Acknowledgements

The development of the Mathematica notebooks described above is supported by the *TEMPUS project 1881/1991*.

References

1. L. J. Segerlind, *Applied Finite Element Analysis*, John Wiley, 1991.
2. B. Szabo and I. Babusks, *Finite Element Analysis*, John Wiley,1991.
3. O. C. Zienkiewicz, K. Morgan, *Finite Element and Approximation*, John Wiley, 1983.
4. E. Hinton and D. R. J. Owen, *Finite Element Programming*, Academic Press, 1977.
5. E. Hinton and D. R. J. Owen, *An Introducttion to Finite Element Computations*, Pineridge Press, 1985.
6. B. Irons and S. Ahmad, *Techniques of Finite Elements*, Ellis Horwood, 1986.
7. St. Wolfram, *Mathematica*, Addison Wesley, 1991.
8. M. D. Mikhailov and M. N. Ozisik, *Heat Transfer Solver*, Prentice Hall, 1991.

IngMath: Software for Engineering Mathematics

H. Bausch
Technical University of Berlin, D-10623 Berlin, Germany

Abstract. The paper is a documentation of the computer program IngMath. This program contains many of the essential numerical methods in a compact form. It is important for the mathematical education of students of technical or other departments. It is very easy to use the program.

Keywords. Computer program, engineering mathematics, numerical methods

1 Applications

The program:
- is helpful for engineers, economists, mathematicians and others to carry out mathematical calculations,
- supports the private studies of students,
- may be used for demonstrations in lectures and seminars.

For instance the program is used in problem seminars "Mathematical modelling of technical problems and solving by mathematical software" for undergraduate students of engineering sciences at the Technical University Berlin.

2 Preferences of the Program

IngMath is a good supplement to other mathematical programs. It is not necessary to read any documentation, so one can think about the mathematical contents only. It is possible to use the program in a complex manner. For instance one can get the numerical solution of a boundary value- or initial value problem and then calculate a spline interpolation function (for this solution). The program allows some possibilities, which are not included in other programs.

3 Short Description of the Program Parts

[ ] Linear algebra:
- *Systems of linear equations*
- *Matrices; Characteristic polynomials, Eigenvalues, Eigenvectors, Inverse matrices.*
[ ] Linear programming:
  Arbitrary linear constraints and nonnegativity conditions.

[ ] Transport optimisation:
   The equilibrium condition "sum of resources = sum of requirements" must not be satisfied.
[ ] Polynomials: For arbitrary polynomials:
- *Calculation of all real and complex roots,*
- *Decomposition into real linear and quadratic factors,*
- *Graphic representation.*
[ ] Rational functions:
- *Decomposition of arbitrary rational functions into partial fractions,*
- *Indefinite integrals,*
- *Definite integrals,*
- *Graphic representation.*
[ ] Functions:
- *Piecewisely defined functions (with an optional parameter) are possible;*
- *Any compositions of 19 basic functions can be used (about these functions see the table in the description of the UNIT),*
- *Roots,*
- *Extrema,*
- *Integration,*
- *Production of datafiles,*
- *Graphic representation.*
[ ] Differential equations:
- *Systems or single equations (with an optional parameter) up to 5th order,*
- *Initial and boundary value problems,*
- *Runge-Kutta method of 4th order with optimisation of steps,*
- *Graphic representation.*
[ ] Spline functions:
- *Natural cubic splines,*
- *Interpolation (Spline and Lagrange),*
- *Fitting (with weighting coefficients),*
- *Data manipulations,*
- *Graphic representation.*
[ ] Non-linear calculations:
- *The calculations are produced with a derivationless direct search method (called PATTERN SEARCH), so one can applicate discontinuous functions and piecewisely defined functions.*
- *Linear/quasilinear/nonlinear regression with free choice of the function and max. 5 regression coefficients, regression by deviance or by sum of absolute differences,*
- *Unconstrained minimisation of nonlinear functions of maximum 5 independent variables (usable also for solving nonlinear equations or systems of equations),*
- *Data manipulations,*
- *Graphic representation.*

[ ] Remarks to the graphic representation:

● *Using the ZOOM function one can get a translation of the picture by 25% of x or y intervals or an increased or decreased picture (by 50% relating to the centre of the picture).*

● *Before you finish a graphic representation you see the picture without any menue. So you are able to print out the "pure" picture. Then go on pressing any key!*

4 Technical Notes

- Volume of the program (full version): approximately 330 KByte.
- Assumed are: AT-computer with a minimum of 2 MByte RAM, DOS, VGA. If the computer has not more than 2 MByte RAM, then the program (full version) can be started under WINDOWS only. Further exists a reduced version 3.28, which needs only 640 KB RAM.
- A coprocessor and a mouse - if exist - are supported.
- The program contains all necessary procedures to print out the numerical results. You can also print out the graphic representations, before loading the DOS - program GRAPHICS (explain the corresponding parameters, if you don't use a matrix printer). Using (for instance) the CLIPBOARD of WINDOWS or the program CAMERA you can save colour pictures and then print out by WINWORD, CORELDRAW *etc.*
- Naturally, because the memory is limited, the program has limited capacities. For instance, the highest possible degree of polynomials is the 20th. It is easy to change that for special cases (contact the author).

The program has been written m PASCAL 7.0 and includes an own formula interpreter (UNIT BAUSCH4), which can be applicated in other programs also and is described m the following.

5 Documentation of the UNIT BAUSCH4

- The interpreter returns the value of an expression (a function), represented by a string (max. length: 254 characters).
- Small and capital letters are allowed.
- A function can have up to 7 independent variables, which must be designated by p,u,v,w,x,y,z. They are defined in the UNIT interface (as REAL). Therefore it is not necessary to define these variables in the program.
- Function values will be calculated with the FUNCTION F defined in the UNIT. For instance, if A is the function string then F(A) returns the function value corresponding to the actual (defined before) values of the independent variables.
- Any compositions of functions may be used (with corresponding signs of aggregation).

- After the input of a function it is to be recommended to examine the Syntax of the expression. For this purpose the UNIT contains a BOOLEAN error function SYNERR. If the syntax of the expression is wrong then SYNERR returns the value TRUE (otherwise FALSE). Use this function as follows: SYNERR(string of all allowed variables, function string) Example: SYNERR('px',A) examines the SYNTAX of the function string A, but the variables x and p are allowed only! The program may fail without examination of the syntax because the interpreter doesn't examine (this would take too much time)!
- After using the function F a BOOLEAN variable ERR can be asked. If the user function is not defined (for the given values of the independent variables), then ERR will return the value TRUE (otherwise FALSE). In such cases the value of F is equal to 0.
- The UNIT version BAUSCH4E uses EXTENDED variables (instead REAL). Sometimes one needs the higher accuracy of these variables.
- The following 19 functions (in any compositions) are allowed:

SIN	SINH	ARCSIN	LN	ABS
COS	COSH	ARCCOS	EXP	SGN
TAN	TANH	ARCTAN	SQRT	BO
COT	COTH	ARCCOT	^INTEGER	

- Numbers may have a maximum of 11 digits. BO is a BOOLEAN function with the value 0 (wrong) or 1 (true). The proposition of BO must contain one and only one symbol <, >, = or symbol combination <=, >=, <>. The function BO allows piecewisely defined functions but also the graphic representation of several functions in one picture.

Examples of allowed functions:

$$x^{\wedge}-3+5*z*\sin((x-y)^{\wedge}2)/(1+\mathrm{sqrt}(\mathrm{arcsin}(w)))$$

$$x^{\wedge}2*\mathrm{bo}(\mathrm{pi}<x)+\tan(x)*\mathrm{bo}(x>=1)+\mathrm{bo}(x>p*\sin(x))$$

$$\sin(x)*\mathrm{bo}(p=1)+\cos(x)*(p=2)$$

The UNIT contains a formula editor also. EDIT(A) reads a string A (max. length: 254 characters) and returns the edited string. Allowed are the same characters as in the formula interpreter (excluding the space key). The usually editor keys can be applicated: All Cursor keys, Backspace, Delete, Home, End. Signs of multiplication (partly the signs of aggregation also) are included automatically.

Example of a Program:

```
USES BAUSCH4;
VAR A: STRING;
BEGIN
A:=";
WRITE('X= '); READLN(X);
REPEAT
WRITE('A= '); EDIT(A);
IF SYNERR('X',A) THEN WRITELN('SYNTAX ERROR!')
ELSE WRITELN(F(A),ERR)
UNTIL NOT SYNERR('X',A);
READLN
END.
```

Power Series Solutions of ODEs

H. Flanders
Department of Mathematics, University of Michigan,
Ann Arbor, MI 48109, USA

Abstract. Automatic differentiation with the use of computer allows one to evaluate the Taylor series of expansion of a function acurately and easily. After showing automatic differentiation of a class of functions, the power series of solutions of ordinary differential equation (ODE) and the system of ODE are explained.

Keywords. Automatic differentiation, Taylor expansion, software, system of ODE

1 Algorithmic Differentiation

We deal with the class of functions $F(t)$ that can be expressed in terms of variables $t, x, y, ...$, real constants, the binary operations $+, -, *, /$, the standard functions of programming languages: sin, cos, arctan, exp, ln, *etc.* What is called *"automatic differentiation"* allows one to evaluate $F(t)$ at a point $t = a$ together with as many derivatives as one pleases. Let us think of the coefficients of the Taylor series expansion of $F(t)$ at $t = a$ rather than the derivatives. Then this computation of a truncated Taylor series is as accurate as the computer, *i.e.*, there is no truncation error, only roundoff error. It may come as a surprise to some that it is possible to calculate the first 100 Taylor series coefficients of

$$F(t) = \tan(\sin t + 5.7 \exp 0.02t)$$

at $t = 3.28$ in about 0.01 sec on a 486/487 class MS-DOS machine.

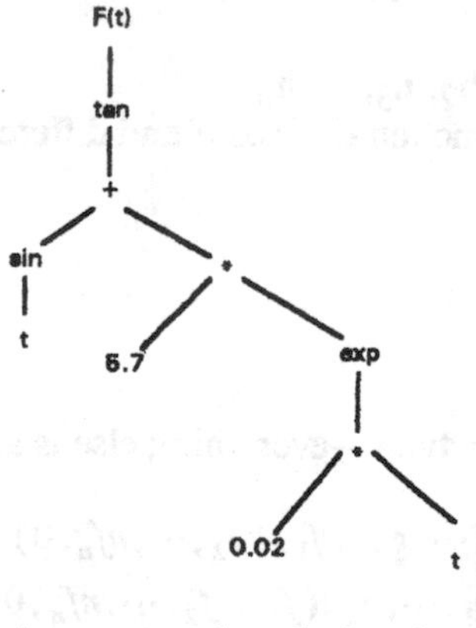

Fig.1 A representation of a function by a binary tree

To see how this works, we must understand how a function can be represented in a computer by a binary tree. A picture of the displayed function should give the idea, Fig 1. Clearly, any function of our class can be represented in this manner.

Suppose that we want to evaluate n derivatives at $t = a$. We attach to each node of the graph a vector $(v_0, v_1, \cdots, v_n)$ of real numbers, representing the values of the derivatives up to that node. To a node containing a constant c we attach the vector $(c, 0, 0, \cdots, 0)$. To each node containing the independent variable t we attach the vector $(a, 1, 0, 0, \cdots, 0)$. If the vectors at all nodes of the graph below a certain node N have been computed, we can compute the vector at N by the rules of differentiation. A few examples will give the idea.

Example 1. $H = F + G$. The vector for H is computed by adding the vectors for F and G.

Example 2. $H = F \cdot G$. Apply the usual formula for polynomial multiplication:

$$h_k = \sum_{i+j=k} f_i g_j .$$

Example 3. $H = \exp F$. Differentiate this relation: $H' = HF'$. In terms of the corresponding vectors:

$$(h_1, 2h_2, 3h_3, \cdots, nh_n, 0) = (h_0, h_1, h_2, \cdots, h_n)(f_1, 2f_2, 3f_3, \cdots, nf_n, 0).$$

Clearly we have $h_0 = \exp f_0$ and the equations

$$h_1 = h_0 f_1 \qquad 2h_2 = 2h_0 f_2 + h_1 f_1 \qquad 3h_3 = 3h_0 f_3 + 2h_1 f_2 + h_2 f_1$$
$$\cdots \quad nh_n = nh_0 f_n + (n-1)h_1 f_{n-1} + \cdots + h_{n-1} f_1$$

These can be solved successively for $h_1, h_2, h_3, \ldots, h_n$.

Example 4. $H = \sin F$. Introduce the function $G = \cos F$ and differentiate:

$$H' = GF' \qquad G' = -HF' .$$

First we have

$$h_0 = \sin f_0 \quad \text{and} \quad g_0 = \cos f_0$$

the only evaluations of transcendental functions—everything else is algebra:

$$(h_1, 2h_2, \cdots, nh_n, 0) = (g_0, g_1, \cdots, g_n)(f_1, 2f_2, \cdots, nf_n, 0)$$
$$(g_1, 2g_2, \cdots, ng_n, 0) = -(h_0, h_1, \cdots, h_n)(f_1, 2f_2, \cdots, nf_n, 0)$$

$$\begin{cases} h_1 = g_0 f_1 \\ g_1 = -h_0 f_1 \end{cases} \quad \begin{cases} 2h_2 = (2g_0 f_2 + g_1 f_1) \\ 2g_2 = -(2h_0 f_2 + h_1 f_1) \end{cases} \quad , \ etc .$$

Each pair of equations can be solved for the next pair of coefficient.

2 Implicit Functions

Suppose a function $x = x(t)$ is defined implicitly by

$$F(t,x) = 0 \qquad F(a,b) = c \qquad F_x(a,b) \neq 0$$

It is possible to compute the Taylor expansion of x automatically by making n pass-pairs through the parse tree of F. Each node t, resp. x has a vector

$$(a,1,0,\cdots,0) \quad \text{resp.} \quad (b,x_1,x_2,\cdots,x_n),$$

where x_1, ..., x_n are unknown. At the root of the parse tree is the node for F, and it contains the vector $(c,0,0,\cdots,0)$.

The zero-th pass through the tree simply produces the known relation $F(a,b) = c$. Pass 1 produces a linear relation $d_1 x_1 + e_1 = 0$, which we solve for x_1. We make a pass through the tree replacing each x_1 at an x node with this value. The second pass determines a new linear relation $d_2 x_2 + e_2 = 0$ which we solve for x_2, then substitute this value, etc.

3 Ordinary Differential Equations (ODE)

The automatic procedure for solving an initial value problem (IVP)

$$\begin{cases} \dfrac{dx}{dt} = \dot{x} = F(t,x) \\ x(t_0) = x_0 \end{cases}$$

is similar to the method for implicit functions. At each node t, resp. x, of the parse tree for F is the vector

$$(t_0,1,0,\cdots,o) \quad \text{resp.} \quad (x_0,x_1,x_2,\cdots,x_n),$$

where t_0 and x_0 are known, but $x_1,\cdots,x_n$ are to be determined. At the root of the parse tree is the vector

$$\dot{x} = (x_1,2x_2,3x_3,\cdots,nx_n,0).$$

The zero-th pass through the parse tree produces the obvious relation

$$x_1 = F(t_0,x_0)$$

and this numerical value is then substituted at each node x. Subsequent passes produce relations

$$2x_2 = c_2, \quad 3x_3 = c_3, \cdots, nx_n = c_n ,$$

where the c's are constants, and these determine all required coefficients of the Taylor expansion of the solution function x.

4 Systems of ODE

The IVP for a system of ODE is handled similarly. Suppose for instance that the system is

$$\begin{cases} \dot{x} = F(t,x,y) & x(t_0) = x_0 \\ \dot{y} = G(t,x,y) & y(t_0) = y_0 \end{cases}$$

Then we construct two parse trees, one each for F and for G. At the root nodes of these trees are the respective vectors

$$\dot{x} = (x_1, 2x_2, 3x_3, \cdots, n\,x_n) \quad \text{and} \quad \dot{y} = (y_1, 2y_2, 3y_3, \cdots, n\,y_n).$$

By traversing the trees simultaneously for each degree, we determine the numerical pairs

$$(x_1, y_1), \quad (x_2, y_2), \quad (x_3, y_3), \cdots, (x_n, y_n)$$

one after the other, and so determine the solution to as high an accuracy as desired.

Implicit systems such as

$$\begin{cases} F(t,x,y,\dot{x},\dot{y}) = 0 \\ G(t,x,y,\dot{x},\dot{y}) = 0 \end{cases} \quad \text{where} \quad \begin{matrix} F(t_0, x_0, y_0, \dot{x}_0, \dot{y}_0) = 0 \\ G(t_0, x_0, y_0, \dot{x}_0, \dot{y}_0) = 0 \end{matrix}$$

can be handled in a similar manner. When we traverse the parse trees simultaneously for degree k, we find a non-singular *linear system* for the pair (x_k, y_k).

5 Software

The following packages of AD (automatic differentiation) software are currently available.

- Chang and G. Corliss, ATOMFT. This is a FORTRAN precompiler. The input is FORTRAN code for a system of ODE and initial data, and the output is a FORTRAN program for solving the system. There are many options. Further information from:*ychang@cmcvx1.claremont.edu* or *george@boris.mscs.mu.edu*
- A. Griewank, *et al*, JAKEF. This is a FORTRAN precompiler for computing gradients and Jacobians of functions of many variables. It is available through NetLib, LibJakef. Further information from: *bischof@mcs.anl.gov* and *juedes @ iastate.edu*.

- O. Garcia, GRAD. Another FORTRAN precompiler for gradients. Available through FTP or email from SIMTEL20/MSDOS.MATH. Further information from: *ogarcia@uchcecvm.bitnet*
- H. Flanders, ODE. A user-friendly, menu-driven program to demonstrate various uses of automatic Taylor series computation. It includes graphics programs for ODE.
  Current status: unfinished in general, and what is finished under alpha test. Will eventually be available commercially, possibly with a manual.

References

[1] Griewank, A., On automatic differentiation. In: Mathematical Programming, Kluwer Academic Press, 1989, 83-108
[2] Griewank, A. and Corliss G. F. (eds), Automatic Differentiation of Algorithms, SIAM Press, 1991
[3] Neidinger, R. D., Automatic differentiation and APL, College Math Journal 20 (1989), 238-250
[4] Rall, L.B., Automatic Differentiation: Techniques and Applications, Lecture Notes in Computer Science 120, Springer Verlag, 1981.

Comments on the Teaching of Computational Fluid Dynamics

A.W. Bush

School of Computing and Mathematics, University of Teesside, UK

Abstract This paper shares the experiences of a skilled practitioner and lecturer in Computational Fluid Dynamics over many years at university level. The topics are restricted to two dimensional fluid flow for incompressible fluids. The author believes that it is desirable for students to use straightforward computer code before tackling more sophisticated software. The paper covers finite difference algorithms, upwinding, primitive variable formulation, and the MAC and SIMPLE algorithms.

Keywords Computational fluid dynamics, Navier-Stokes equations, stream function, primitive variables, hovercraft model.

1 Introduction

I am going to describe some features concerning the teaching of computational fluid dynamics (CFD) for two dimensional, incompressible fluids. These are governed by the Navier-Stokes equations:

$$\rho(\frac{\partial u}{\partial t} + \frac{u\partial u}{\partial x} + \frac{v\partial u}{dy}) = -\frac{\partial p}{\partial x} + \mu\nabla^2 u + b_x \qquad (1)$$

$$\rho(\frac{\partial v}{\partial t} + \frac{u\partial v}{\partial x} + \frac{v\partial v}{dy}) = -\frac{\partial p}{\partial y} + \mu\nabla^2 v + b_y \qquad (2)$$

and the continuity equation which for constant density becomes

$$\frac{\partial u}{\partial x} + \frac{\partial v}{\partial y} = 0 \qquad (3)$$

There are very few exact solutions of these equations. The use of numerical solutions greatly enlarges the range of flows which can be analysed. In my experience, fluid dynamics courses are enriched by the inclusion of some computer solutions. There are various CFD packages available and some are reasonably priced for academic use on small machines.

It is desirable that students have experience of using a straightforward computer code before they use one of the more sophisticated packages. The straightforward code can either be programmed by the students themselves or presented to them for their modification to treat various flows described by implementing appropriate boundary conditions.

There are two common approaches to CFD namely the use of the stream function/vorticity formulation and the use of primative variables.

2 The Stream Function/ Vorticity Formulation

The stream function, ψ, is introduced such that

$$u = \frac{\partial \psi}{\partial y} , v = - \frac{\partial \psi}{\partial x} \tag{4}$$

then any velocity field obtained from a stream function ψ will automatically satisfy the incompressibility constraint (3).

The pressure is eliminated from the Navier Stokes equations by subtracting the y derivative (1) from the x derivative of (2) this leads to the occurrence of the vorticity

$$\omega = \frac{-\partial u}{\partial y} + \frac{\partial v}{\partial x}$$

(This is the z component of curl $\underline{v}$ and for 2D flow it is the only non zero component of curl $\underline{v}$).

The result of eliminating the pressure from the Navier Strokes equations leads to the vorticity transport equation,

$$\rho \left(\frac{\partial \omega}{\partial t} + \frac{u\partial \omega}{\partial x} + \frac{v\partial \omega}{\partial y} \right) = \mu \nabla^2 \omega + \frac{\partial b_y}{\partial x} - \frac{\partial b_x}{\partial y} \tag{5}$$

Introducing the stream function derivatives for the velocity components in the definition of vorticity leads to the following equation.

$$\nabla^2 \psi = -\omega \tag{6}$$

Equations 5 and 6 provide two equations from which ω and ψ may be obtained. The non linear term in the material derivative namely,

$$(\underline{v}.\underline{\nabla})\omega = \frac{\partial \psi}{\partial y} \frac{\partial \omega}{\partial x} - \frac{\partial \psi}{\partial x} \frac{\partial \omega}{\partial y}$$

requires that an iterative procedure be used to obtain a solution.   A straightforward scheme is to use an explicit formula for ω from (5),

$$\frac{\omega_{ij}^n - \omega_{ij}^o}{Dt} = \text{terms involving } \omega^o \text{ and } \psi^o \text{ at } i, j \text{ and } i \pm 1, j \pm 1 \quad (7)$$

Here the superscript n and o denote new and old time levels respectively   while the i and j subscripts denote spacial locations where $x = iDx$ and $y = jDy$. The spacial derivatives are approximated using standard finite difference approximations. Thus for example

$$\nabla^2 \omega \simeq \frac{\omega_{i+1j} - 2\,\omega_{ij} + \omega_{i-1j}}{Dx^2} + \frac{\omega_{ij+1} - 2\omega_{ij} + \omega_{ij-1}}{Dy^2} \quad (8)$$

The time step is included even if a steady state solution is sort. The solution proceeding either from an initial condition or an initial estimate of a desired steady flow.

Having obtained the new time level or steady state estimate of ω^n this is   used as a source term in Poissons equation (6) for ψ^n.

In many flows the convection terms are dominant and it is found that "upwinding" these terms allows larger time steps to be used whilst maintaining a stable scheme.

The upwinding scheme may be described with reference to the term $u\frac{\partial\omega}{\partial x}$. This is approximated as follows,

$$u\frac{\partial\omega}{\partial x} = \begin{cases} u\,(\omega_{ij} - \omega_{i-1j})/Dx, & u > 0 \\ u\,(\omega_{i+1j} - \omega_{ij})/Dx, & u < 0 \end{cases}$$

There are larger truncation errors associated with this procedure than with central differences. This gives rise to the occurrence of so called numerical diffusion [1]. This can be overcome by the use of higher order differencing.

If a steady state solution is sort the iterative proceedure can be made more stable whilst maintaining an explicit scheme by replacing ω_{ij}^o wherever it occurs on the right hand side of (7) with ω_{ij}^n and taking those terms onto the left hand side of (7). Thus with upwinding, the coefficient of ω_{ij}^n instead of being 1/DT becomes,

$$1/Dt + |u| / Dx + |v| /Dy + 2\nu\,(1/Dx^2 + 1/Dy^2).$$

This increase in the coefficient of the unknown ω_{ij}^n leads to increased stability allowing larger values of Dt to be chosen.

If a transient solution is to be determined then larger time steps can only be used in conjunction with an implicit scheme.

The above algorithm determines ψ and ω values at interior points of a finite difference mesh. Their values on the boundary require a different procedure depending on the nature of the boundary conditions. In the case of fully contained flow the boundary will be a curve of constant ψ value. This is conveniently chosen to be zero. The vorticity is then obtained by imposing equation 6 on the boundary. This requires an approximation for $\nabla^2\psi$ on the boundary. This is obtained using fictitious values of ψ outside the flow region. In figure 2.1 below the case of an upper boundary moving tangentially with speed U_o is considered.

$\bullet\psi_N$ (fictitious)

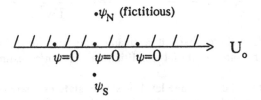

$\psi=0 \quad \psi=0 \quad \psi=0$ U_o

ψ_S

Fig 2.1 - A Moving Boundary

Then $\omega = -\nabla^2\psi = -\left(\dfrac{\psi_N + \psi_S}{Dy^2}\right)$ where $(\psi_N - \psi_S)/\,2Dy = U_o$

hence ω is determined from ψ_S. Fixed boundaries simply have U_o equal to zero.

At an inlet where the normal velocity is prescribed the stream function varies across the inlet so as to create this flow. Thus for a uniform inflow between $y = 0$ and $y = h$ for example with $u = U_o$ we have $\frac{\partial\psi}{\partial y} = U_o$ for $0 < y < h$, thus $\psi = \psi_o + U_o y$ (see fig 2.2).

$\psi_o + U_o h$

$y=h$

$U_o \longrightarrow$

$y=0$

ψ_o

Fig 2.2 - A Uniform Inflow

Similarly at outlets if the velocity profile is known then the stream function is obtained by integration. However it is more usual at outlets for the pressure to be prescribed. This type of boundary condition is difficult to implement using the ψ/ω formulation. Indeed this is one of the reasons why the primitive variable formulation is advantageous.

Another difficulty with the ψ/ω formulation concerns interior solid objects on whose surface the stream function is difficult to determine.

Further disadvantages of the formulation are:
- Restricted to 2D
- Need for a pressure recovery calculation
- Turbulence is best described in the primitive variable formulation.

3 The Primitive Variable Formulation

In this approach the velocity and pressure fields are solved for directly.   The Navier Stokes equations provide natural equations for the velocity components in an iterative solution procedure. The continuity equation remains to determine the pressure. This leads to an immediate difficulty since the pressure does not appear in equation (3). One way around this difficulty is to introduce an artificial compressibility relating a pressure change to the divergence of the velocity field. This is called the penalty formulation [2].

The algorithm used in many commercial CFD codes is a pressure correction scheme derived by imposing the continuity equation in such a way as to obtain a Poisson type equation for the pressure.

The easiest of such schemes to describe and implement is the simplified Marker and Cell algorithm. The original MAC code was developed by Harlow and Welch [3] to deal with free surfaces. This involved the introduction of marker particles which followed the motion of the surface. In the case of flows without free surfaces it is not necessary to introduce marker particles. The simplified scheme consists of the following stages.

(i)   Initial values of the velocity u^o, v^o and pressure field p^o are provided or, in the case when a steady flow is sort, initial values are estimated. (The time dependence is preserved even for steady flows so as to generate the iterative scheme).

(ii)   Intermediate velocity fields u^* and v^* are obtained from the following explicit form of the Navier-Stokes equations,

$$\frac{u^*-u^o}{Dt} = -u^o \frac{\partial u^o}{\partial x} -v^o \frac{\partial u^o}{\partial y} + \nu \nabla^2 u^o + b_x - \frac{1}{\rho} \frac{\partial p^o}{\partial x} \qquad (9)$$

$$\frac{v^*-v^o}{Dt} = -u^o \frac{\partial v^o}{\partial x} -v^o \frac{\partial v^o}{\partial y} + \nu \nabla^2 v^o + b_y - \frac{1}{\rho} \frac{\partial p^o}{\partial y} \qquad (10)$$

At this stage the velocity field (u*, v*) will not necessarily satisfy the continuity condition.

(iii)   The pressure corection equation

$$\nabla^2 q = \frac{\rho}{Dt} \left(\frac{\partial u^*}{\partial x} + \frac{\partial v^*}{\partial y} \right) \qquad (11)$$

is solved for q and the pressure corrected using

$$p^n = p^o + q \qquad (12)$$

Equation (11) is derived below.

(iv)   The new velocity field (u^n, v^n) are obtained from the pressure correction as follows,

$$\frac{u^n - u^*}{Dt} = -\frac{1}{\rho} \frac{\partial q}{\partial x} \qquad (13)$$

$$\frac{v^n - v^*}{Dt} = -\frac{1}{\rho} \frac{\partial q}{\partial y} \qquad (14)$$

(v)   Update u^o, v^o and p^o and return to stage (i).

Equation (11) is derived by taking the combination $\frac{\partial}{\partial x} (13) + \frac{\partial}{\partial y} (14)$ and imposing $\nabla \cdot \underline{v}^n = 0$, the pressure correction is then governed by Poissons equation with the divergence of $\underline{v}^*$ as the source term.

The MAC scheme described above requires a time step sequence. It is possible to obtain a pressure correction scheme for a steady state by using an implicit solution procedure. The SIMPLE scheme (Patankar and Spalding [4]) is based on this idea and forms the basis of a number of commercial codes.

The space derivatives in equations (9) - (14) are approximated by finite difference equations. It is helpful to introduce a staggered computational grid where the pressure is evaluated at the cell centres and the normal velocities on the cell sides as shown in fig 3.1.

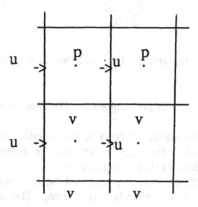

Fig 3.1   The Staggered Grid

The staggered grid has the following advantages:

-    Governing equations involve the flux of material from one cell to another. The velocity on the cell sides ensures that the mass leaving one cell enters another.

-    Pressures are not required on boundary walls which are made to coincide with the cell boundaries.

-    The occurence of the so called chequerboard instability is avoided. In 1D this is described by considering a grid where the velocity and pressure are evaluated at the same points.

p	p	p	p+c	p	p+c
$\vec{u}$	$\vec{u}$	$\vec{u}$	$\vec{u}$	$\vec{u}$	$\vec{u}$

Fig 3.2   The Chequerboard Instability

In the second case shown in fig 3.2 the addition of any value c to alternate pressures will not be seen by the Navier Stokes equations which involve first order pressure derivatives

$$\frac{\partial p}{\partial x} = \frac{P_E - P_W}{2Dx}$$

In the case of the staggered grid no such oscillations occur. (See [5]).

   Boundary conditions are easily imposed. Common examples are described below.
   At boundaries where the normal velocity is known either to be zero on a   wall or a prescribed inlet value then $\underline{v}.\underline{n}$ is set to the same value for the old, intermediate and new values so that dq/dn is zero.
   At boundaries where the pressure is prescribed the pressure correction will be zero. This is a common outlet boundary condition. The continuity equation is here not required for the calculation of q and is available to determine the normal component of the outlet velocity field.
   The tangential velocity boundary condition is obtained either from a   prescribed shear, often set to zero at an inlet or outlet, or the no slip boundary condition at a solid wall. In this case the momentum equations 9 and 10 require fictitious velocity fields at half a grid space outside the flow region. This is determined so that the value of the fluid velocity interpolated at the wall is that of the solid wall (see fig 3.3).

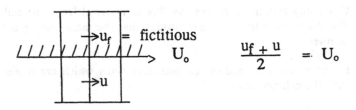

Fig 3.3   The No Slip Boundary

4 Results

Results are shown in Fig 4.1 for the driven cavity flow where the motion of the top wall induces recirculating flow by the action of the viscous shear force. Results are shown in Fig 4.2 for a hovercraft model where an inlet flow is specified along with an exit reference pressure. The pressure exerted on the top right hand boundary surface corresponds to the lift field by the hovercraft.

It is of interest to vary the wall speed, U_0, dimension, L, and viscosity, ν, in the driven cavity flow. The location of the vortex point moves to the right as the Reynolds number, $U_0 L/\nu$ increases.

In the hovercraft model it is of interest to reduce the size of the outlet gap and note the increase in the lift force generated by the flow.

The results shown for both flows have a Reynolds number of 100. A 20x20 grid was used with a time step of 0.005. Non dimensional velocities v/U_0 and pressures $p/\rho U_0^2$ are presented where U_0 is the wall speed for the cavity and the inlet speed for the hovercraft.

References

1. Roache, P.J.: Computational fluid dynamics. Albuquerque NM: Hermosa 1972
2. Peyret, R., Taylor, T.D.: Computational Methods for Fluid Flow, Berlin: Springer 1983
3. Harlow, F.H., Welch, J.E.: Numerical calculation of time dependent incompressible flow of a fluid with a free surface. Phys Fluids 8, 2182-2189 (1965)
4. Patankar, S.V., Spalding, D.B.: A calculation procedure for momentum transfer in three-dimensional parabolic flows. Int J Heat and Mass Transfer 15, 1787-1806 (1972)
5. Patanker, S.V.: Numerical heat transfer and fluid flow. Hemisphere 1980.

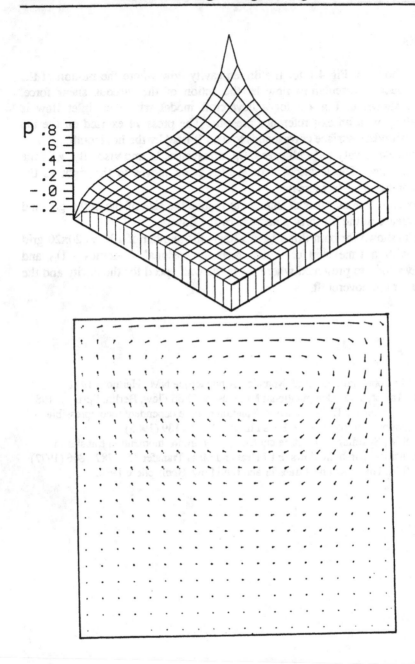

Fig 4.1   Pressure and Velocity Fields for Driven Cavity Flow

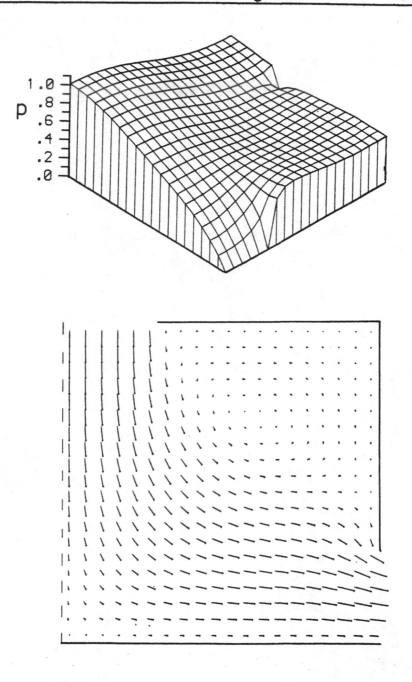

Fig 4.2 Pressure and Velocity Fields for the Hovercraft Model

PART THREE

Mathematical Modelling in Fluid Flow and Heat Transfer

This part contains two major papers: one by Professor Kakaç on Fluid Flow and one by Professor Yener on Heat Transfer. These papers, by experts in their fields, demonstrate the finished product, i.e., fully working models which are used by engineers.

Model One aims to predict low frequency oscillations in boiling systems for single-channel, horizontal, two-phase flow. Experiments with various heat inputs are described and compared with model predictions and the comparisons are good.

Model Two applies spectral methods to the numerical solution of heat transfer problems. In particular an example involving mixed convection in a horizontal two-dimensional channel is discussed in detail. This paper also contains a good literature survey of fundamental instabilities which trigger transition from laminar to turbulent flow.

Finite Difference Modelling of Two-Phase Flow Instabilities in Boiling Systems

S. Kakaç and V. R. Mullur
University of Miami, Coral Gables, FL 33124, USA

Abstarct. Previous experimental work in forced convection boiling channels has suggested the presence of both high frequency (density-wave type) and low frequency (pressure-drop type) oscillations. This paper aims to predict the low frequency oscillations in a system using R-11 as the working fluid, working on the hypothesis that the instabilities are "system compressible volume" related. Experimental results shown are obtained from a setup at the University of Miami. The modeling is done for the setup in consideration. The steady state system pressure-drop characteristics are determined by a numerical solution of the governing equations as derived from the Drift-flux models. The transient characteristics of the flow are obtained by perturbing the governing equations around the steady state values obtained. Oscillations were obtained for various heat inputs, flow rates, and exit restriction diameters. Both the steady-state and transient solutions are obtained using an Explicit Finite Difference scheme. Good agreement between the theory and experiments is obtained.

Keywords. Boiling system, two-phase flow, flow instability, finite difference modelling, drift-flux model, oscillation, transient solution

1 Introduction

Two phase flows have been observed to occur in many industrial situations like refrigeration systems, turbo-machinery, and power plant heat exchangers. Prediction of flow parameters such as steady-state pressure-drop characteristics, stability boundaries during boiling, and oscillations is very crucial in the design of two-phase flow equipments. Thermal oscillations are undesirable as they can cause problems of system control, and can lead to tube failure due to wall temperature increase. They can also lead to fatigue failure of the tube caused by a continuous cycling of the wall temperature. It is clear from these problems that plant equipments should posses an adequate margin of safety against oscillations.

This paper aims to theoretically study the steady-state and oscillatory characteristics of single channel horizontal two-phase flows. The drift-flux model is used and its results are compared with the experimental findings.

2 Theoretical Study

In the most general formulation of the two-phase flow problem, the conservation equations are written separately for each of the phases, hence it is called as a "separated flow model". Various forms of the conservation equations have been proposed in the literature [1-4]. In most of the practical problems, however, one dimensional time-dependent equations are used. To close the set of six conservation equations (three for each phase), seven constitutive laws are required friction and heat transfer for the two phases at the tube wall, and three conservation equations for shear force, mass balance and energy balance at the interface of the two phases. The requirement of seven constitutive laws makes the use of this model very difficult. In order to reduce the complexity involved in the formulation of the problem in its most general form, as noted above, several models have been suggested, which attempt at correlating different parameters of the two-phases, e.g. drift velocity, void fraction, slip ratio, etc. In the solution of two-phase problems by one these correlations, the six phase equations are written separately and then combined for the mixture [2]. For a number of two-phase flow regimes, such as annular or slug flow, the homogeneous model does not reflect the physics of the phenomena, since the assumption of equality of the phase velocities is not justifiable. The drift-flux formulation which has gained much acclaim in the last decade, takes the relative velocity between the phases into account, while assuming thermodynamic equilibrium [5]. In addition to the assumption used in the development of the homogeneous model, the pressure and other fluid properties are assumed to be uniform along the cross-section. The effect of the radial distribution of the void fraction is represented by a distribution parameter, C_0. The energy equation is limited to a thermal balance over a control volume. In the boiling region, the fluid flow is treated as a mixture of saturated liquid and vapour phases travelling at different velocities. The relative velocity between the phases is represented by the so-called drift velocity, u_{vj}, of the vapour phase with respect to the centre-of-volume of the mixture.

2.1 Mechanism of Dynamic Instabilities

Before we study the mathematical formulation of the instability problem it is useful to look at the mechanism of some observed two-phase flow instabilities [6,7].

2.1.1 Pressure-drop Oscillations.   With reference to Fig.1a, the following steady-state relations are considered

$$(P_I - P_s) = K_1 G_i^2. \tag{1}$$
$$(P_s - P_e) = \psi(G_o). \tag{2}$$

Here P_I is the main tank pressure, P_s the surge tank pressure, P_e the exit pressure; K_1 is an experimentally determined constant for the inlet restriction, G_i is the mass velocity into the surge tank and G_o is the mass velocity out of the surge tank. The first equation represents the pressure drop across the inlet restriction and is a statement of the momentum equation across the restriction. Note that the surge tank is downstream of the restriction. The second equation is the pressure drop between the surge tank and the system exit section, and it is the system curve shown in Fig. 1.

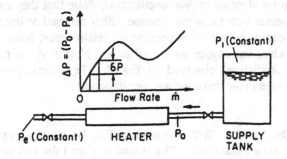

Fig. 1a Simplified system for pressure-drop type oscillations

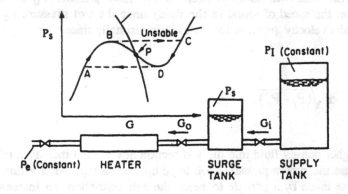

Fig. 1b Simplified system for density-wave type oscillations

This equation is incorporated as a steady state numerical algorithm in the computer program. While operating on the negative slope portion, a slight increase in the surge tank pressure would cause more fluid to enter it than leaves it. The surge tank pressure now increases due to accumulation of fluid. The operating point moves up until it reaches the peak (point B). Any higher pressure can be sustained only by a higher mass flow rate as given by the system curve. This point is found to be in the single phase liquid region (point C). At C, the amount of fluid leaving the surge tank is more than the amount entering it. Therefore, the surge tank pressure decreases till the operating point reaches the curve minima at D. Any lower pressure can now be obtained only if the mass flow rate reduces to the value at point A. Hence, an excursion to A is observed. Now the mass leaving the surge tank is lesser than that enters it. Hence pressure goes up pushing up the operating point till A is reached, where once again a flow excursion is observed. Thus perturbation at any point on the negative slope region results in a flow oscillation tracing the limit cycle ABCDA. This is essentially the mechanism of pressure drop oscillations. Note that they can occur only if the system possesses a compressible volume, either external or internal. In actual heat exchange equipments there is present at certain times, some internal compressible volume which can trigger such oscillations. Note a typical recording of the pressure-drop oscillations observed in Fig. 2. The superimposed high frequency disturbance seen is the density-wave oscillation.

2.1.2 Density-wave Oscillations. With reference to Fig. 1b, we now explain the mechanism of density-wave oscillations. The pressure at the inlet and exit reservoirs are kept constant at all times. It is assumed that the rate of vapour generation in the test section is constant at all times. Suppose, that at time t=0, the exit restriction pressure drop undergoes a sudden infinitesimal drop from its steady state value. This will cause a drop in the inlet pressure P_0 almost instantaneously(at the speed of sound in the fluid) since the exit pressure P_e is constant. The inlet velocity (u_i) now increases infinitesimally since

$$u_i \, \alpha \, \sqrt{(P_I - P_o)}. \qquad (3)$$

This sends a higher density fluid into the test section at t=0. An increased inlet velocity will cause the system pressure drop to go up (see curve). After a time t, which is the time taken by a particle to reach the exit restriction, an increased pressure drop causes the inlet pressure P_0 to increase since the exit pressure is constant. An increased inlet pressure causes a decrease in inlet velocity (note above equation). A decreased inlet velocity in turn causes the residence time of the particle to go up and thus the fluid has greater enthalpy (and lesser density) when it reaches the restriction.

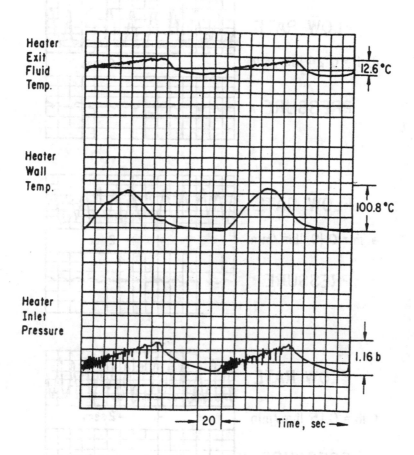

Fig. 2 Typical recordings of superimposed pressure-drop type oscillations; Heat Input: 500 W, Inlet Temperature: -0.3⁰C, Mass Flow Rate: 0.0099 kg/s [8]

A lower inlet velocity causes a lesser pressure drop at the restriction and this starts the cycle again. From the preceding analysis it is evident that it takes around twice the residence time of a fluid particle for a complete set of events to repeat. Density-wave type oscillations observed during experimentation is presented in Fig. 3.

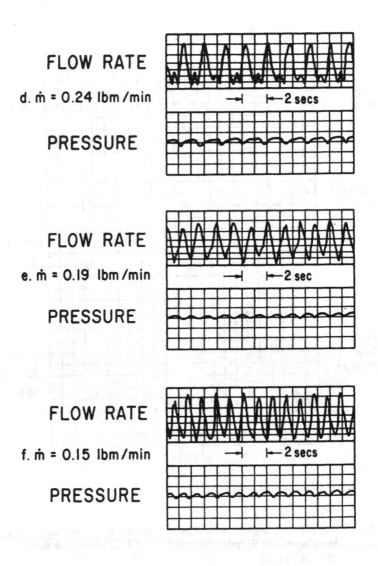

FLOW RATE

d. $\dot{m}$ = 0.24 lbm /min

PRESSURE

FLOW RATE

e. $\dot{m}$ = 0.19 lbm /min

PRESSURE

FLOW RATE

f. $\dot{m}$ = 0.15 lbm /min

PRESSURE

Fig. 3 Typical recording of density-wave type oscillations;
1090Btu/Hr, Inlet Temperature: 25^0 F [9]

2.2 Mathematical Formulation of the Drift-Flux Model

The following equations are written for a horizontal, single channel up flow boiling system.

2.2.1 *Conservation Equations* The continuity, energy, and momentum equations are as follows:

Continuity equation:

$$\frac{\partial}{\partial t}[\rho_1(1-\psi)+\rho_v\psi]+\frac{\partial}{\partial z}[\rho_1 u_1(1-\psi)+\rho_v u_v\psi]=0. \quad (4)$$

Here Ψ is the local void fraction or the volumetric concentration. The above equation is actually a sum of two equations, one each for the liquid and vapour phases, as can be seen from the weighting parameter Ψ.

Energy equation:

$$\frac{\partial}{\partial t}[\rho_1(h_1 - Pv_1)(1-\psi)+\rho_v(h_v - Pv_v)\psi]$$

$$+\frac{\partial}{\partial z}[\rho_1 u_1 h_1(1-\psi)+\rho_v u_v h_v\psi]=\frac{q'}{AL_h}, \quad (5)$$

where ϕ is the heat input to the fluid per unit volume of the fluid, h_1 and h_v, the enthalpies of the liquid and vapour phases, and A, the area of cross section of the flow duct. Note here that the energy equation is basically a thermal balance over a part of the heated section. The right hand side term denotes the heat flux input. The first term on the left hand side is the rate of change of total (liquid + vapour) internal energy within the control volume and the second term in the net enthalpy convected into the control volume

Momentum equation:

$$-\frac{\partial P}{\partial z} = \frac{\partial}{\partial t}(\rho_1 u_1(1-\psi) + \rho_v u_v \psi)$$
$$+ \frac{\partial}{\partial z}[G^2[\frac{(1-x)^2}{\rho_1(1-\psi)} + \frac{x^2}{\rho v \psi}]] + (\frac{\partial P}{\partial z})_{tp,fric}$$

(6)

The first and second terms of the right hand side of the equation denote the momentum build-up due to fluid flow and two phase frictional drop respectively. The pressure drop is a function of the wall roughness and flow regime and is generally an empirical correlation in terms of the Reynolds number. The last term on the right hand side of the equation is the net efflux of momentum due to vapour and liquid flows. x is the quality in the above equation. The mass flux, G, in the momentum equation is defined as

$$G = \rho_1 u_1 (1-\psi) + \rho_v u_v \psi.$$

(7)

The phase velocities, u_1 and u_v, will be related to each other using the drift-flux kinematic constitutive relation derived by Zuber and Findlay [5].

2.2.2 Auxiliary Terms The various auxiliary terms in the governing equations are defined below.

Mass velocity, G,

$$G = \rho_1 u_1(1-\psi) + \rho_v u_v \psi.$$

(8)

Quality (dryness fraction), x,

$$x = \frac{\rho_v u_v A_v}{\rho_v u_v A_v + \rho_1 u_1 A_1}.$$

(9)

Void Fraction, ψ,

$$\psi = \frac{A_v}{A}.$$

(10)

Total Enthalpy at a given section,

$$h = \rho_1 h_1 (1 - \psi) + \rho_v h_v \psi \qquad (11)$$

2.2.3 Equations of State The thermo-physical properties of R 11 have been cor-
related in a polynomial form from data available in [11]. The saturated properties
are in terms of pressure and are valid in the range of 0.6 to 6.2 Bar. Enthalpies
units are kJ/kg (specific enthalpy), and density is expressed in kg/m^3.

Saturated Liquid Enthalpy:

$$h_f = 20.11198 + 42.14375p - 10.8984p^2$$
$$+ 1.659071p^3 - 0.09865p^4 \qquad (12)$$

Saturated Vapour Enthalpy:

$$h_v = 215.8316 + 24.78123p - 6.6406p^2$$
$$+ 1.009815p^3 - 0.06008p^4 \qquad (13)$$

Saturated Liquid Density:

$$\rho_1(p) = 1489.7 - 0.11p + 0.0473p^2 \qquad (14)$$

Saturated Vapour Density:

$$\rho_v(p) = 0.233297 + 5.798266p - 0.23475p^2$$
$$+ 0.036673p^3 - 0.00204p^4 \qquad (15)$$

2.2.4 *Two-phase Frictional Pressure Drop* To estimate the magnitude of the
two-phase frictional pressure gradient, a two-phase friction multiplier F_M is used
along with an expression for the single-phase pressure drop. F_M is defined as

$$F_M = \frac{\Delta P_{TP}}{\Delta P_{SP}} . \qquad (16)$$

This factor is a function of the quality of the two-phase flow. Some researchers correlate the two-phase factor in terms of the mixture viscosity (the quality being implicit in the relation). The single-phase pressure gradient can be estimated by using the definition of the Fanning friction coefficient as

$$(\frac{dP}{dz})_{sp,fric} = 4\frac{f_o}{d}\frac{G^2}{2\rho_1}, \tag{17}$$

where f_o is the single phase friction factor. It can be estimated by using Blasius' formula as

$$f_o = 0.079(\frac{Gd}{\mu_1})^{-n}, \tag{18}$$

where n is taken as 0.25. Thus the two phase pressure frictional pressure gradient takes the form

$$(\frac{dP}{dz})_{TP} = 4\frac{f_o}{d}\frac{G^2}{2\rho_1}F_M. \tag{19}$$

2.2.5 The Kinematic Correlation for Void Fraction To solve the three conservation equations presented in the previous sections, we need to find relationships between the phase velocities in terms of the volumetric flow rates and void fraction. The model to be used in the present study was originally proposed by Zuber and Findlay [5]. A volumetric flux for the two-phase mixture is defined as

$$j = u_1(1-\psi)+u_v\psi. \tag{20}$$

Equating the mass flux of the liquid at any section to the liquid fraction of the total mass velocity, we get,

$$\rho_1 u_1(1-\psi) = G(1-x) \tag{21}$$

Similarly for the vapour mass flux, we can write,

$$\rho_v u_v \psi = Gx \tag{22}$$

Thus, the volumetric flux j can be expressed in terms of the mass velocity, G, as

$$j = \frac{G(1-x)}{\rho_1} + \frac{xG}{\rho_v} \quad . \tag{23}$$

According to the Zuber-Findlay model, the vapour velocity, u_v, may be related to the volumetric flux as,

$$u_v = C_o\, j + u_{vj} \, , \tag{24}$$

where C_0 is the distribution parameter and u_{vj} is the drift velocity of the vapour phase with respect to the centre of mass of the mixture. In the literature, there are various correlations for C_0 and u_{vj}, depending primarily on the flow pattern. The following expressions used in the present study are reported to give good results for high pressure steam-water flows irrespective of the flow pattern [5].

$$C_o = 1.13 \tag{25}$$

$$u_{vj} = 1.41[\frac{\sigma g(\rho_1 - \rho_v)}{\rho_1^2}]^{1/9} \tag{26}$$

3 Two-phase Flow Characteristics-Solution Procedure

The study of two-phase flow dynamic instabilities, in general, requires the knowledge of the steady-state pressure-drop versus mass flow rate characteristics, over the range of interest. The stability boundaries for pressure-drop and density-wave oscillations are usually shown on the plot of these relationships. These relationships, which are the steady-state solutions to the conservation equations, are also used to determine the initial conditions for both type of oscillations. Therefore, initially solutions are obtained for various heat inputs and/or inlet subcoolings under steady-state conditions. The heater inlet temperature is taken to be $24^{\circ}C$.

3.1 Steady-State Characteristics

Under steady-state conditions, the time-dependent terms in the governing Eqs. (4), (5), and (6) drop out and the following equations are obtained:

Continuity equation:

$$\frac{\partial}{\partial z}[\rho_1 u_1(1-\psi) + \rho_v u_v \psi] = 0. \tag{27}$$

Energy equation:

$$\frac{q'}{A} = \frac{\partial}{\partial z}[\rho_1 u_1 h_1 l(1-\psi) + \rho_{,} u_{,} h_{,} \psi]. \tag{28}$$

Momentum equation:

$$-\frac{\partial P}{\partial z} = \frac{\partial}{\partial z}[G^2 (\frac{(1-x)^2}{\rho_1 (1-\psi)} + \frac{x^2}{\rho_{,} \psi})]$$

$$+(\frac{\partial P}{\partial z})_{fric} + [\rho_1 (1-\psi) + \rho_{,} \psi] \tag{29}$$

3.1.1 Region-wise Finite-difference Equations The momentum and energy equations are to be integrated over the system. Due to the pressure dependence of the properties, it is necessary to numerically perform the integrations. All fluid properties are expressed in terms of pressure. In writing the finite-difference equations, five distinct regions are identified along the system, each having different characteristics, Figs. 4 and 5. For each of these regions finite-difference formulations of the flow are made as follows:

Region 1 (Upstream tubing): The flow is incompressible and adiabatic. This whole region is taken as one lump interval. Therefore the conservation equations for steady-state reduce to:

Continuity:

$$\frac{\partial G}{\partial z} = 0. \tag{30}$$

Energy:

$$\frac{\partial}{\partial z}[\rho_1 u_1 h_1 (1-\psi) + \rho_{,} u_{,} h_{,} \psi] = 0. \tag{31}$$

Momentum:

$$\frac{\partial P}{\partial z} = -\frac{\partial P}{\partial z}\big|_{fric} - \rho_1 g. \tag{32}$$

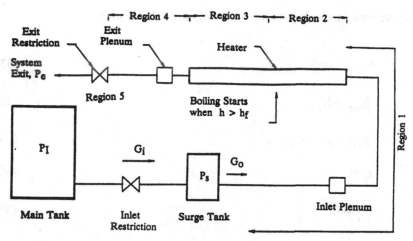

Fig. 4 Schematic diagram of the model for finite-difference analysis;
Horizontal flow channel

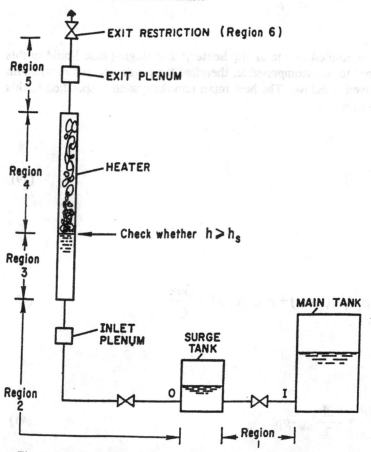

Fig. 5 Schematic diagram of the model for finite-difference analysis;
Vertical flow system

The above equations reduce to the following discretized forms:

$$G_{i+1} = G_i = G.$$ (33)

$$h_{i+1} = h_i.$$ (34)

$$\rho_{i+1} = \rho_i = \rho_1$$ (35)

$$x_i = 0$$ (36)

$$\psi_i = 0$$ (37)

$$P_{i+1} = P_i - 2f\frac{\Delta z}{d}\frac{G^2}{\rho}\Big|_i - \rho_1(\Delta z).$$ (38)

Region 2 (The subcooled region of the heater): The single-phase liquid in this region is assumed to be incompressible, therefore the continuity and momentum equations are given as below. The heat input into the system is specified in this region, *i.e.*, ϕ = constant.

Continuity:

$$\frac{\partial G}{\partial z} = 0.$$ (39)

Energy:

$$\frac{\partial}{\partial z}[\rho_1 u_1 h_1(1-\psi) + \rho_v u_v h_v \psi] = \frac{\phi\Delta z}{G}.$$ (40)

Momentum:

$$\frac{\partial P}{\partial z} = -\frac{\partial P}{\partial z}\Big|_{fric} - \rho_1 g.$$ (41)

The above equations reduce to the following discretised forms:

$$G_{i+1}=G_i=G \qquad (42)$$

$$h_{i+1} = h_i + \frac{q'\Delta z}{GA_h} \qquad (43)$$

$$\rho_{i+1} = \rho_i = \rho_1. \qquad (44)$$
$$x_i = 0. \qquad (45)$$
$$\psi_i = 0. \qquad (46)$$

$$P_{i+1} = P_i - 2f\frac{\Delta z}{d}\frac{G^2}{\rho}\Big|_i - \rho_i g(\Delta z). \qquad (47)$$

Region 3 *(The boiling region of the heater):* The conservation equations for this region are below:

Continuity:

$$\frac{\partial G}{\partial z} = 0 . \qquad (48)$$

Energy:

$$\frac{\partial}{\partial z}[\rho_1 u_1 h_1 (1-\psi) + \rho_, u_, h_, \psi] = \frac{q'}{A} \qquad (49)$$

Momentum:

$$\frac{\partial P}{\partial z} = -2\frac{f}{d}\rho u^2 - \frac{\partial}{\partial z}[G^2 (\frac{(1-x)^2}{\rho_1(1-\psi)} + \frac{x^2}{\rho_,\psi})] \qquad (50)$$
$$- g[\rho_1(1-\psi) + \rho_,\psi]$$

These equations are discretised as follows:

$$G_{i+1}=G_i=G \qquad (51)$$

$$h_{i+1} = h_i + \frac{\phi \partial z}{\rho u} \qquad (52)$$

$$P_{i+1} = P_i - 2f\frac{\Delta z}{d}\frac{G^2}{\rho^i} - G^2\left(\left[\frac{(1-x_{i+1})^2}{\rho_l(1-\psi_{i+1})} + \frac{x_{i+1}^2}{\rho_v \psi_{i+1}}\right]\right.$$

$$\left.-\left[\frac{(1-x_i)^2}{\rho_l(1-\psi_i)} + \frac{x_i^2}{\rho_v \psi_i}\right]\right) - g\Delta z[\rho_l(1-\psi_{i+1}) + \rho_v \psi_{i+1}]$$

$$(53)$$

The equations are discretised as follows:
The quality, x, is found out as:

$$x_{i+1} = \frac{h_{i+1} - h_{f,i+1}}{h_{fg,i+1}}$$

$$(54)$$

The average volumetric flux is obtained by

$$<j>_{i+1} = G\left[\frac{(1-x_{i+1})}{\rho_{l,i+1}} + \frac{x_{i+1}}{\rho_{v,i+1}}\right].$$

$$(55)$$

The average void fraction $<\Psi>_{i+1}$ is written as:

$$<\psi>_{i+1} = \frac{x_{i+1}G}{\rho_{v,i+1}<j>_{i+1}}\left[1.2 + \frac{1.53}{<j>_{i+1}}\frac{\sigma g\rho_{i+1}}{\rho_{l,i+1}^2}\right]^{1/4}\right]^{-1}$$

$$(56)$$

Region 4 (After heater):

(a) Before the exit restriction: There is no heat input to the fluid at this section, thus the enthalpy is a constant. Density does not vary and the only source of pressure drop is the wall friction drop. This section can therefore be treated as a lump and the equations are similar to that of Region 1.

Continuity:

$$\frac{\partial G}{\partial z} = 0 \quad .$$

$$(57)$$

Energy:

$$\frac{\partial}{\partial z}[\rho_l u_l h_l(1-\psi) + \rho_v u_v h_v \psi] = 0.$$

$$(58)$$

Momentum:

$$\frac{\partial P}{\partial z} = -\frac{\partial P}{\partial z}\Big|_{fric} - \rho_i g.$$ (59)

The above equations reduce to the following discretised forms:

$$G_{i+1} = G_i = G$$ (60)

$$h_{i+1} = h_i$$ (61)

$$\rho_{i+1} = \rho_i$$ (62)

$$x_{i+1} = x_i$$ (63)

$$\psi_{i+1} = \psi_i$$ (64)

$$P_{i+1} = P_i - 2f\frac{\Delta z}{d}\frac{G^2}{\rho_i} - \rho_i g(\Delta z)_o.$$ (65)

(b) Exit restriction: Two-phase flow is observed at this section. The exit restriction is a sharp-edged orifice of diameter, 2.64 mm. An empirical correlation, based on experimental data, is used to calculate the pressure-drop across the restriction. The two-phase pressure drop across the restriction is found from an expression of the form

$$\Delta P_e = \Delta P_{SP} F_M,$$ (66)

where ΔP_{SP} is the single-phase pressure drop across the orifice plate. This is experimentally determined as

$$Dp_{SP} = 175\frac{G^2}{\rho_1}.$$ (67)

The two-phase multiplier F_M too, has been experimentally determined as

$$F_M = 1 + 28.73x_e - 6.68x_e^2 + 25.518x_e^3,$$ (68)

where x_e is the quality of the liquid vapour mixture at the exit of the heated section.

3.1.2 Boundary Conditions. The conservation equations, together with the equations of state and the constitutive relations, are to be solved for the following boundary conditions:

- constant inlet temperature, T_i = constant,
- constant heat input, ϕ = constant, and,
- constant exit pressure, P_e = constant.

3.1.3 *Scheme of Solution.* Calculations start with given values of mass velocity G, inlet fluid temperature T_i, heat input ϕ, and exit pressure P_e. Assuming an inlet (surge tank) pressure P_s, the flow parameters and properties are calculated from the exit of the surge tank to the inlet of the heater. The enthalpy, pressure and density are calculated at each successive node in the heater. At each step, the enthalpy is checked against the saturated enthalpy at that pressure. Boiling is assumed to start when the fluid enthalpy exceeds the saturation liquid enthalpy. Appropriate state and constitutive equations are chosen according to the state of the fluid. The calculation is continued along the rest of the system. The calculated exit pressure is checked against the given value of P_e. If the difference is found to be within an acceptable margin, the assumed value of surge tank pressure P_s is substituted for the surge tank pressure; otherwise the whole procedure is repeated with a different value of P_s, till convergence is obtained. This scheme is represented in the flow-chart given in Fig. 6.

3.2 Time-Dependent Solutions-Pressure-Drop Oscillations

Using the method and results developed in the previous sections, the time dependent behaviour of the system under consideration is predicted for pressure-drop type oscillations.

3.2.1 *Model for Pressure-drop Type Oscillations.* Pressure-drop oscillations that occur in two-phase flow systems are triggered by a small instability in the negative slope region of the steady state characteristics curve. The surge tank is an important dynamic component of the system that serves as an "external compressible volume". These oscillations have relatively low frequencies, and their periods are usually much larger than the residence time of single fluid particle in the system. Consequently, the transient operating points can be obtained as a series of steady state points. For the surge tank, the continuity equation can be written as

$$\frac{dP_s}{dt} = P_s^2 \frac{(G_i - G_o)A_p}{P_{so}V_o\rho_1}. \tag{69}$$

The momentum equation for the mass velocity, G_i, between the main tank and the surge tank is written as:

$$G_i = [\frac{(P_I - P_s)\rho_1}{K_I}]^{\frac{1}{2}} \tag{70}$$

where K_I is the inlet restriction coefficient of the valve between the main and the surge tanks. The above expression has been experimentally determined as

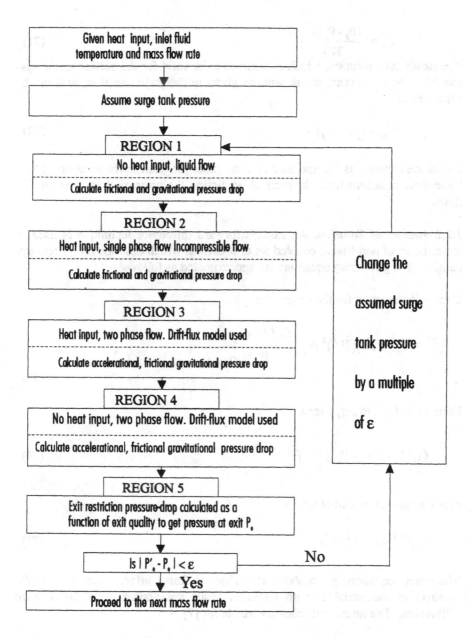

Fig. 6 Flow chart for the steady-state algorithm

$$G_i = [\frac{(P_I - P_s)\rho_1}{328}]^{1/2}.$$ (71)

The steady state solution, which is assumed to be valid for the section of the system after the surge tank, is the same as given in the earlier section, and can be expressed as:

$$G_o = G_o(P_s, Q_I).$$ (72)

The above equation is incorporated in a computer program in the solution. Thus, these three equations form the basis of predicting the pressure-drop type oscillations.

3.2.2 Method of Solution. An explicit forward difference technique is used to solve the set of non-linear, coupled set of equations which describe the system dynamics. The governing equations are approximated as follows:

Continuity equation for the surge tank:

$$P_s^{j+1} = P_s^{j} + \Delta t[(P_s^{j})^2 A_p \frac{(G_i^{j+1} - G_o \, j+1)}{P_{so} V_o \rho_1}].$$

(73)

From main tank to surge tank:

$$G_i^{j+1} = [(P_I - P_s^{j})\frac{\rho_1}{K_I}]^{\frac{1}{2}}.$$ (74)

From surge tank to exit of system:

$$G_o^{j+1} = G_o(P_s^{j}, Q_I^{j}).$$ (75)

The system equations given above are written in finite difference form. The derivatives of the variables with respect to time are approximated by forward differences. The numerical scheme is stable for [7]

$$\Delta t^{j+1} \leq \frac{\Delta z}{Max|u_i^{j}|}.$$ (76)

The calculations start with given fluid temperature and pressure. The initial flow parameters and properties, corresponding to the given inlet mass velocity and heat input, are calculated using the steady-state program. These results are saved as the initial conditions at the stable operating point.

The system is perturbed by increasing the pressure P_S in the surge tank. The inlet mass velocity for the surge tank is calculated using Eq. (73). Calling the steady-state program again, the mass velocity and other flow parameters and properties along the system are computed using Eq. (74), corresponding to the increased surge tank pressure. Since the pressure in the surge tank is increased, the flow rate through the heater decreases. The heat transfer coefficient and the heat input into the fluid also change. Using Eq. (72), the pressure P_S in the surge tank is calculated. As the exit mass velocity G_O from the surge tank decreases more than the mass velocity G_i into the surge tank, P_S increases further during the time step Δt. This procedure is repeated in the successive time steps. This scheme is represented in the flow-chart given in Fig. 7. In the successive time steps, the pressure difference between the surge tank and the system exit increases along the negative slope, until the top of one of the steady-state mass flow rate versus pressure-drop curves is reached, depending on the instantaneous heat input into the fluid. Then the calculation point leaps from the top of the curve to the liquid region automatically. The flow rate through the heater increases. The surge tank pressure decreases with increasing flow rate through the heater, since the exit flow rate is larger than the inlet flow rate. The pressure in the surge tank decreases till the bottom of one of the steady-state mass flow rate versus pressure-drop curves. Then another flow excursion takes place from the liquid region to two-phase or vapour region. These limit cycles are then repeated. In the following chapter, the results obtained from the modelling are presented along with the obtained experimental data. A comparison of the results of the drift-flux and homogeneous flow models is also presented.

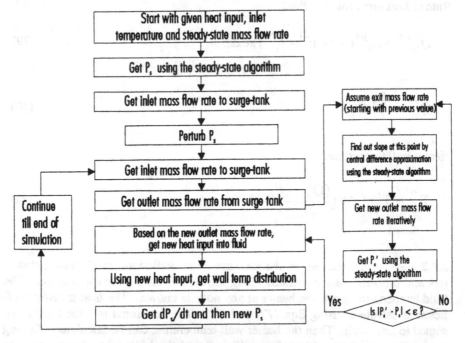

Fig.7 Flow chart for the time-dependent solution algorithm

3.3 Time-dependent Solutions-Thermal Oscillations

During the pressure-drop oscillations, the mass flow rate, heat transfer coefficient, and heat input into the fluid keep changing. However the heat generated in the heater wall is constant. Therefore, when the limit cycle enters the liquid region, the wall temperature decreases as the liquid heat transfer coefficient is usually high. Thus the wall temperature fluctuates during the limit cycle. These are called thermal oscillations [12,13].

3.3.1 *Model for Thermal Oscillations.* The rate of heat transfer into the fluid is given by:

$$Q_{I,i} = \pi \, d \, \Delta z \, \alpha_i (T_{w,i} - T \, f,i). \tag{77}$$

The heater wall temperature can be calculated from the energy balance for the heater, yielding,

$$\frac{d(T_w)}{dt} = \frac{(Q_o - Q_I)}{m_h c_h}. \tag{78}$$

The finite-difference form of the governing equations, Eqs. (77) and (78), is written as follows:

Rate of heat input into the fluid:

$$Q_{I,i}{}^{j+1} = \alpha_i{}^{j+1} (T \, w,i^{j+1} - T_f{}^{j+1}) \pi d \Delta z \tag{79}$$

and

$$Q_I{}^{j+1} = \sum_{i=1}^{n} Q_{I,i}{}^{j+1}. \tag{80}$$

Heater wall temperature:

$$T_{w,i}{}^{j+1} = T_{w,i}{}^{j} + \Delta t [(\frac{(Q_o - Q^j{}_{I,i})}{(MC)_h}]. \tag{81}$$

3.3.2 *Method of Solution.* In the pressure-drop oscillation model, fluid parameters and properties are calculated along the system during the oscillations. The fluid temperature inside the heater at any node is known. The heat transfer coefficient is calculated using Eqs. (77) and (78). The heat input into the fluid is assumed to start with. Then the heater wall temperature can be calculated. During the oscillations, the heat transfer coefficient and the heat input change, and the heater wall temperature changes accordingly. The governing equations are written in forward finite difference form. The solution of these yield the thermal oscillations at any node of the heater.

4 Comparison of Theoretical and Experimental Results

We shall now examine the results from both the theoretical study and experimental work performed on vertical flow and horizontal flow systems.

4.1 Steady-State results

Figs. 8 and 9 and Table 1 show the pressure-drop versus mass flow rate results for a constant inlet fluid temperature ($T_i = 20^0$ C), with various electrical heat input rates (Q_0= 0-2500 W). The theoretical predictions for both vertical and horizontal flow systems can be seen to be in very close agreement with the experimental results over the entire range of parameters involved. The region of negative slope between the regions of positive slope plays a critical role in generating and sustaining the oscillations. The experiments also indicate that as the heat input rate increases, the magnitude of the negative slope increases, thus making the system more unstable. These important features can be seen to be predicted very well by the theoretical model.

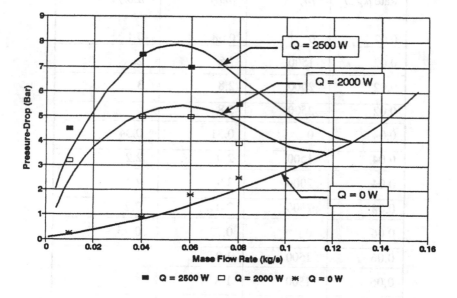

Fig. 8 Horizontal two-phase flow steady-state characteristics; Comparison of
drift-flux model and experimental results. The exit restriction diameter
is 2.64 mm and the tube diameter is 10.90 mm. Working Fluid: R-11

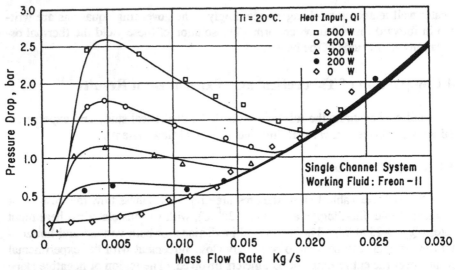

Fig. 9 Vertical two-phase flow steady-state characteristics; Comparison of drift-flux model and experimental results. The exit restriction diameter is 2.64 mm and the heater tube diameter, O.D. is 12.50 mm, I.D. is 11.30 mm. Working Fluid: R-11

Mass Flow Rate *(kg/s)*	Heat Input *(W)*	Pressure-Drop *(Bar)*	Pressure-Drop *(Bar)*
0.02	0	0.08	0.1
0.02	1500	1.8	2
0.02	2000	2.8	3
0.02	2500	5.8	6
0.04	0	0.31	0.34
0.04	1500	2.0	2.2
0.04	2000	3.9	4.1
0.04	2500	7.0	7.2
0.06	0	0.73	0.75
0.06	1500	1.8	2
0.06	1500	1.8	2
0.06	2000	3	5

Table 1. Pressure-drop and mass flow rate for steady state flow

4.2 Time-Dependent Results

Figs. 10 through 13 and Tables 2 and 3 show the pressure-drop and thermal oscillations for both vertical and horizontal flow systems. It is seen that the pressure-drop type and thermal oscillations are slightly out of phase with each other. The rising portion of the pressure-drop oscillations corresponds to an increasing vapour flow that carries away more heat, thus lowering the wall temperature. The decreasing portion corresponds to a decreasing liquid mass flow rate that convects progressively lower heat away from the wall. This causes the temperature to increase. The experimental and theoretical results shown exhibit this characteristic clearly and are themselves in good agreement. The experimental cases are selected so that at least two different cases of relevant parameters are represented, *i.e.* operating mass flow rate, heat input, inlet subcooling, and the heater surface condition. It can be seen that there is good agreement between the two. The periods, amplitudes, as well as the waveforms, of the oscillations are reasonably well predicted by the theory. Tables 1 and 2 summarise the comparison between the experimental and theoretical results. The pressure-drop oscillations result through the interaction between the flow and the compressible volume in the surge-tank. Under the present experimental conditions, high frequency density-wave oscillations also occur, which are superimposed on the pressure-drop oscillations. The present model can predict the pressure-drop oscillations quite well. However, density-wave oscillations cannot be predicted by this model, because it does not take into account explicitly the propagation of continuity waves that generate these oscillations. Efforts are under way to extend the modelling methodology in order to predict the complete superimposed oscillations.

5 Conclusions

Experiments with various heat inputs have been conducted using nichrome tubes in a single channel up flow boiling system and horizontal system to study the pressure-drop type and thermal oscillations. Experiments have been performed at a constant inlet temperature for different heat inputs. The drift-flux formulation of two-phase flows has been developed to obtain the steady-state characteristics of the vertical channel system. The following conclusions can be reached based on the experimental and theoretical studies:

- Both the pressure-drop type and thermal oscillations occur at all heat inputs. At a given inlet subcooling, the amplitudes and periods of the oscillations increase with increasing heat input rate.

- Both the pressure-drop type and thermal oscillations occur at all inlet subcoolings. At a given heat input rate, the amplitudes and periods of the oscillations increase with increasing inlet subcooling.

- Thermal oscillations accompany the pressure-drop type oscillations. Oscillations of pressure and temperature are in phase; but the maximum of temperature oscillations always lag the maximum of pressure oscillations.

- The periods and amplitudes of the pressure-drop type oscillations increase with decreasing mass flow rate at the initial operating point on the negative slope.

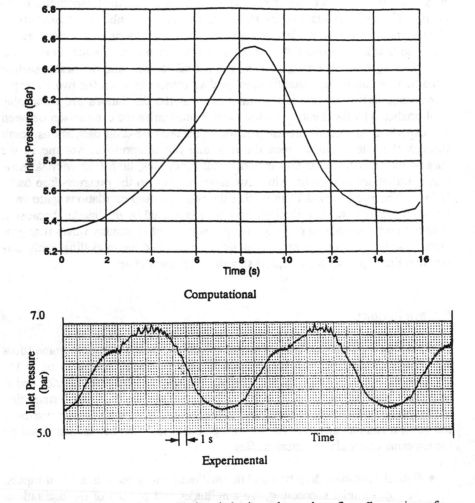

Computational

Experimental

Fig. 10 Pressure-drop oscillations in horizontal two-phase flow. Comparison of theoretical model and experimental results. The exit restriction diameter is 3.175 mm and the tube diameter is 10.90 mm. The heat input to the fluid is 2500 W. Working Fluid: R-11, Mass Flow Rate: 0.0717 kg/s

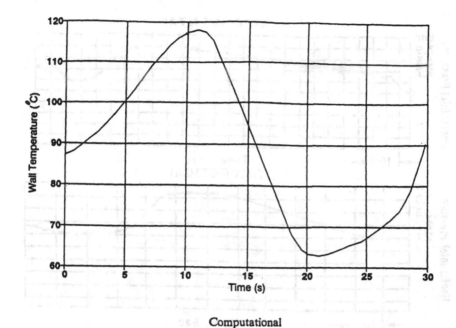

Computational

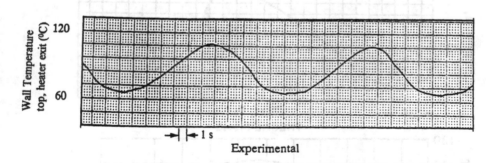

Experimental

Fig. 11 Thermal oscillations in horizontal two-phase flow. Comparison of
theoretical model and experimental results. The exit restriction diameter
is 2.64 mm and the tube diameter is 10.90 mm. The heat input to the fluid
is 2500 W. Working Fluid: R-11, Mass Flow Rate: 0.0717 kg/s

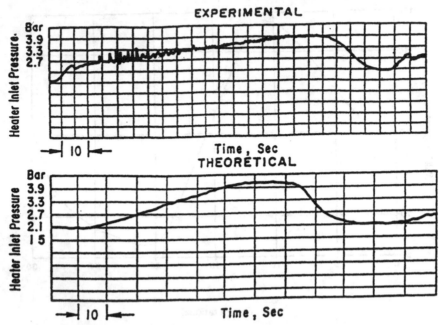

Fig. 12 Pressure-drop oscillations in vertical two-phase flow. Comparison of
theoretical model and experimental results. The exit restriction diameter
is 3.175 mm and the tube diameter is 10.90 mm. The heat input to the fluid
is 500 W. Working Fluid: R-11, Mass Flow Rate: 0.005242 kg/s

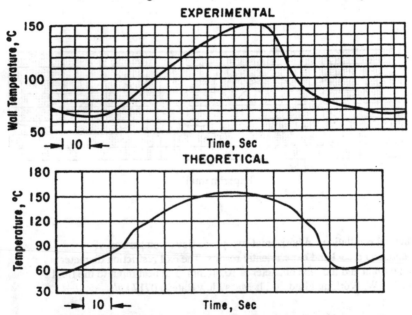

Fig. 13 Thermal oscillations in vertical two-phase flow. Comparison of theoretical
model and experimental results. The inlet temperature is - 8° C, the heater
tube outer diameter is 12.5 mm , inner diameter is 11.3 mm, Working
Fluid: R-11, Mass Flow Rate: 0.00542 kg/s

Exit Restriction (mm)	Heat Input (W)	Experimental		Theoretical	
		Period (s)	Amplitude (Bar)	Period (s)	Amplitude (Bar)
2.64	2000	18	.1.5	24	1.9
2.64	2500	16	2.2	18	2.0
3.175	2000	13	1.1	12	1.25
3.175	2500	14	1.2	12	1.4

Table 2. Comparison between theoretical and experimental results

Heat Input (W)	Mass Flow Rate (kg/s)	Oscilation Type	Period (s)	Amplitude	Period (s)	Amplitude
800	7.31	Heater Inlet Pressure	50	0.96 Bar	57	0.94 Bar
		Heater Exit Pressure	50	99.8° C	57	96.0° C
600	7.31	Heater Inlet Pressure	25	0.50 Bar	30	0.69 Bar
		Heater Exit Pressure	25	49.2° C	30	40.0° C
800	11.89	Heater Inlet Pressure	28	0.89 Bar	27	1.14 Bar
		Heater Exit Pressure	28	65° C	27	75.0° C

Table 3. Mass flow rate and pressure-drop oscillations

- The steady-state characteristics and the oscillations predicted with the use of the use of the drift-flux model are in reasonably good agreement with experimental results.

Acknowledgements. The authors gratefully acknowledge the partial finacial support of the NATO Scientific Affairs Division. The authors also wish to extend their sincere appreciation to Ms. Renhua Shen, Visiting Scholar at the University of Miami, for her help during the preparation of the final draft of this paper.

Nomenclature

A   Tube inner surface area, m^2

A_p Surge tank cross-section area, m^2

d   Inner diameter of the heater tube, m

f   Friction factor, dimensionless

F_M Two-phase flow friction multiplier, dimensionless

g   Gravitational acceleration, $9.806\ m^2/s$

G   Fluid mass velocity ($= \rho u$), $kg/(m^2\ s)$

G_i Inlet mass velocity to surge tank, $kg/(m^2/s)$

G_o Outlet mass velocity from surge tank, $kg/(m^2/s)$

h   Specific enthalpy of the fluid, J/kg

h_{lv} Latent heat of vaporization, J/kg

j   Volumetric flux, m/s

K_I Inlet restriction coefficient, dimensionless

P   Pressure, Pa

P_o System inlet pressure during density-wave type oscillations, Pa

P_{so} Steady-state pressure in surge tank, Pa

Q_I Heat input into the fluid during oscillations, W

t   Time, s

T   Fluid inlet temperature, °C

u   Fluid velocity, m/s

V_o Volume of noncondensable gas in surge tank, m^3

x   Quality of the liquid-vapor mixture, dimensionless

z   Axial distance along the flow path, m

Greek Symbols

μ Dynamic viscosity of the fluid, Pa s

ρ Density, kg/m^3

σ Surface tension, N/m

ϕ Heat input to fluid per unit volume of fluid, W/m^3

Ψ Void Fraction, dimensionless

Δ Incremental length along tube

Subscripts

e   Exit

f   Fluid

i   Inlet

l   Liquid

o   Heater exit to restriction

s   Surge tank

v   Vapor

w   Wall

References

[1] Ishii M.: Drift-flux model and derivation of kinematic constitutive laws. In: S. Kakaç, F. Mayinger and T.N. Veziroglu *(eds)*, Two-phase Flows and Heat Transfer, Hemisphere, New York (1977)

[2] Boure, J.A., Bergles, A.E. and Tong, L.S.: Review of two-phase flow instabilities, Nuc. Engg & Design 25, 165-192 (1973)

[3] Padki, M.M.: Theoretical Modeling of Two-phase Flow Instabilities in a Vertical Upflow Boiling System, Ph.D. dissertation, Uni of Miami, Coral Gables, Florida, (1990)

[4] Yadigaroglu, G., and Lahey, R.T., Jr.: On the various forms of conservations equations in two-phase flow, Int. J. of Multiphase Flow 2, 477-484 (1976)

[5] Zuber, N. and Findlay, J.A.: Average volumetric concentration in two-phase flow systems, ASME J. Heat Transfer 87, 453-468 (1965)

[6] Stenning, A.H.: Instabilities in the flow of a boiling liquid, J. Basic Eng., Trans. ASME, Series D 86, 213-217 (1964)

[7] Kakaç, S., Liu, H.T.: Two-phase flow dynamic instabilities in boiling systems. In: X.J Chen *et al (eds)*, Multiphase Flow and Heat Transfer 1, Hemisphere Publ., Washington (1991)

[8] Veziroglu, T.N. and Kakaç, S.: Two-Phase Flow Instabilities, Final Report, NSF Project CME 79-20018 (1983)

[9] Akyuzlu, K. *et al*: Finite difference analysis of two-phase flow pressure-drop and density-wave type oscillations, Warme-und Stoffubertragung 14, 253-267 (1980)

[10] Padki, M.M., Liu, H.T., and Kakaç, S.: Two-phase flow pressure-drop type and thermal oscillations, Int. J. Heat and Fluid Flow 12(3), 240- 248 (1991)

[11] Irvine Jr., T.F., Hartnett, J.P.: Steam and Air Tables in SI Units, Hemisphere Publ., Washington (1976)

[12] Kakaç, S., *et al* : Investigation of thermal instabilities in a forced convection upward boiling system, Int. J. Experimental Thermal and Fluid Scie 3, 191-201 (1990)

[13] Kakaç, S., *et al:* Two-phase flow thermal instabilities in a vertical boiling channel, ASME Winter Annual Meeting, Dec 1988, Chicago. In: E.E. Michaelides and M.P. Sharma *(eds)* Fundamentals of Gas-Liquid Flows 72, ASME, FED (1988).

Numerical Modelling in Heat Transfer by Spectral Methods

M. Zenouzi[1], Y. Yener[2] and A. Tangborn[2]

[1]School of Technology, Youngstown University, OH 44555, USA
[2]DME, Northeastern University, Boston, MA 02115, USA

Abstract. Spectral methods belong to general class of weighted residual techniques which approximate continuous functions globally in terms of a Fourier or polynomial series expansion. For problems with sufficiently smooth solutions, spectral methods provide exponential convergence. The most commonly used series expansions are the Fourier series for periodic problems and Chebyshev and Legendre polynomials series for non-periodic problems. The choice of the basis functions and the manner of computing the expansion coefficients together characterize the method (e.g., Galerkin, Tau, or collocation). The objective here is to discuss the implementation of spectral methods to the numerical solution of heat transfer problems with examples.

Keywords. Spectral method, numerical modelling, heat transfer, discretization, Chebyshev-Tau method, collocation method

1 Introduction

Spectral methods (and spectral element methods) in computational heat transfer studies have proven successful in a large number of problems involving complex physical phenomena in simple, as well as in complex geometries. For instance, in the areas of stability, transition, turbulence, and mixed convection the resolution and accuracy afforded by spectral techniques have allowed for the simulation and, therefore, understanding of highly complicated and delicate flow and heat transfer phenomena [1].

Spectral methods are one of the major numerical simulation techniques to approximate solutions of partial differential equations. They belong to the general class of weighted residual techniques [2, 3], which permit to define approximations of functions in terms of a truncated Fourier or polynomial series expansion with an error term which is forced to be zero in a weighted average sense. Unlike finite-difference methods, a numerical solution obtained by using a spectral method is defined everywhere in the domain, not just at grid points. More specifically, spectral methods provide rapid (exponential) convergence (the error decreases exponentially with increasing resolution; whereas, in a second- order accurate finite-difference method, the error decreases only as the square of the spatial resolution) and are, therefore, more economical than finite–differences. For problems with sufficiently smooth solutions, spectral methods require fewer degrees of freedom (grid points or modes) per spatial dimension than second-order finite–differences, and therefore storage requirements are greatly reduced compared to finite–difference or finite–element techniques. The details of the material presented in this section can be found in References [1, 4 – 8].

As an example, Peyret [8] compared the accuracy of various numerical methods for the solution of the one-dimensional Helmholtz equation

$$-\nu\frac{d^2u}{dx^2} + u = 1, \qquad -1 < x < 1, \tag{1}$$

with the boundary conditions

$$u(-1) = 1, \qquad u(1) = 0. \tag{2a, b}$$

The exact solution of this problem is given by

$$u(x) = 1 - \frac{\sinh[(x+1)\sqrt{1/\nu}]}{\sinh(2\sqrt{1/\nu})}. \tag{3}$$

For the numerical solutions the following methods were employed:
1. Second-order finite-difference method.
2. Fourth-order compact method.
3. Tau-Chebyshev method.
4. Collocation-Chebyshev method.

In methods 1 and 2, a variable mesh of Gauss-Lobatto-Chebyshev type was employed. For $\nu = 0.01$ he presented results for the RMS error defined as

$$\bar{E} = \left[\frac{1}{N-1}\sum_{i=1}^{N-1}|u_N(x_i) - u(x_i)|^2\right]^{\frac{1}{2}} \tag{4}$$

at the collocation points for different values of N. The results are shown in Fig. 1, which clearly demonstrates the accuracy of spectral methods.

Fourier spectral methods can be used successfully for periodic problems, *i.e.*, for the cases where the physical problem naturally inhabits a periodic domain, like ocean forecasting, coastal modeling, current-system simulation, and global atmospheric flow models used by meteorologists for routine weather forecasting. Indeed, spectral models are now used routinely by most national weather centers (see Bourke [9] for more details).

In the following sections we describe the basics of the spectral methods, and briefly review the Galerkin, Tau, and collocation methods. Next, we give the implementation of collocation methods, and discuss the collocation points y_i and the Lagrangian interpolants $h_i(y)$ for the Chebyshev and Legendre collocation methods. Finally, in the last section we present differentiation matrices

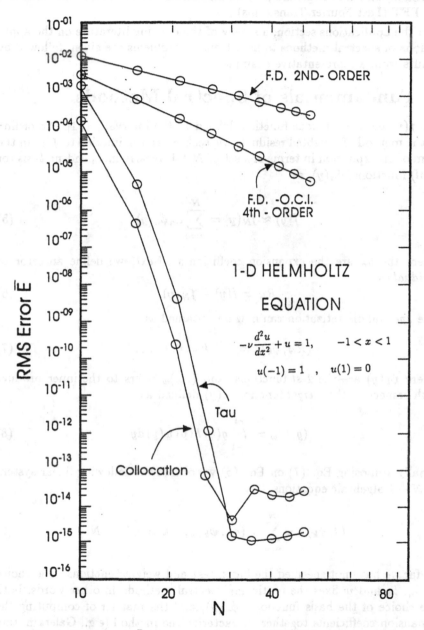

Fig. 1. RMS error $\bar{E}$ for the solution of 1-D Helmholtz equation [8].

for Chebyshev and Legendre polynomials and also Chebyshev differentiation via FFT (Fast Fourier Transforms).

In the applications section, a survey of the existing literature on the applications of spectral methods to heat transfer problems are given, followed by results from a representative example.

2. Fundamentals of Spectral Methods

Let $f(y)$ be a continuous function defined in the interval [-1, 1]. According to the method of weighted residuals, we seek an approximation to $f(y)$ in the form of an expansion in terms of a set of $N+1$ linearly independent *basis* (or *trial*) functions, $\phi_n(y)$, as

$$f(y) \approx f_N(y) = \sum_{n=0}^{N} a_n \phi_n(y) \tag{5}$$

where the a_n are the expansion coefficients. Next, we define an error or *residual* as

$$R_N = f(y) - f_N(y). \tag{6}$$

The *spectral* discretization can now be obtained as

$$(R_N, \psi_k)_w = 0, \quad k = 0, 1, ..., N, \tag{7}$$

where $\psi_k(y)$ are the *test* functions, and $(.,.)_w$ refers to the inner product with respect to the *weight* function $w(y)$ defined as

$$(g, h)_w = \int_{-1}^{1} g(y)\, h(y)\, w(y)\, dy. \tag{8}$$

Finally, imposing Eq. (7) on Eq. (6) we obtain the following linear system of $N+1$ algebraic equations:

$$(f, \psi_k)_w = \sum_{n=0}^{N} a_n (\phi_n, \psi_k)_w, \quad k = 0, 1, ..., N. \tag{9}$$

So far we have not specified the basis, test and weight functions. The choice of ϕ_n, ψ_k and w fixes the particular spectral method. In other words, both the choice of the basis functions, $\phi_n(y)$, and the manner of computing the expansion coefficients together characterize the method (e.g., Galerkin, tau, or collocation).

We now describe below these methods briefly with greater emphasis on the collocation methods. The reader is referred to Canuto *et al.* [4], Gottlieb and Orszag [6], Peyret [8], and Solomonoff and Turkel [10] for further details and more advanced treatments of various aspects of spectral methods.

2.1 Galerkin Method

In this method, the basis functions belong to a set of orthogonal functions, that is,

$$(\phi_m, \phi_n)_w = c_n \delta_{mn}, \tag{10}$$

where δ_{mn} represents the Kronecker–delta and c_n is the norm of ϕ_n given by

$$c_n = (\phi_n, \phi_n)_w. \tag{11}$$

The weight function w is associated with the orthogonality of the basis functions. Also, the basis and test functions are chosen to be the same, that is,

$$\psi_n(y) = \phi_n(y), \quad n = 0, 1, ..., N, \tag{12}$$

and each basis function $\phi_n(y)$ satisfies the same (homogeneous) boundary conditions as the function f being approximated. Thus, the system of linear algebraic equations given by Eq. (9) reduces to

$$a_n = \frac{1}{c_n}(f, \phi_n)_w, \quad n = 0, 1, ..., N. \tag{13}$$

The choice of $\phi_n(y) = e^{iny}$ with $w = 1$ results in a Fourier–Galerkin method. The choice of Legendre polynomials for ϕ_n with $w = 1$ gives the Legendre–Galerkin method, and similarly the choice of Chebyshev polynomials with $w = (1 - y^2)^{-1/2}$ results in the Chebyshev–Galerkin method.

2.2 Tau Method

The tau method is a modified form of the Galerkin method allowing the use of basis functions which do not satisfy the same homogeneous boundary conditions as $f(y)$ or if $f(y)$ has inhomogeneous boundary conditions. Similar to the Galerkin method, the $N + 1$ expansion coefficients, a_n, are obtained by considering

$$(R_N, \phi_n)_w = 0, \quad n = 0, 1, ..., N - m, \tag{14}$$

where m is the number of boundary conditions. The missing m equations are obtained by the application of the boundary conditions.

2.3 Collocation Method

In this method, which is also called the pseudospectral method or the method of selected points, the test functions are chosen as

$$\psi_k(y) = \delta(y - y_k), \quad k = 0, 1, ..., N, \tag{15}$$

where δ represents the Dirac delta-function and y_k are the prescribed collocation (selected) points in (-1, 1). In addition, the weight function is taken to be unity, that is,

$$w(y) = 1. \tag{16}$$

Thus, the discretization given by Eq. (7) reduces to

$$R_N(y_k) = 0, \quad k = 0, 1, ..., N, \tag{17a}$$

or

$$f(y_k) = \sum_{n=1}^{N} a_n \phi_n(y_k), \quad k = 0, 1, ..., N. \tag{17b}$$

In the collocation method the residual is forced to be exactly zero at certain points (collocation points), while in the Galerkin method the residual is zero in an average sense. In this method, any differentiation and quadrature are done with spectral approximations, and all multiplications and divisions are done on a grid of points. The representation of the function goes back and forth between spectral and physical spaces by use of discrete (and fast) transforms. The use of collocation methods simplifies the treatment of various boundary conditions and coordinate transformations considerably. It is especially convenient for differential equations with variable coefficients or for nonlinear differential equations.

The system given by Eq. (17b) is a discrete Fourier expansion. This system can be inverted by means of a discrete orthogonality property when the set $\{y_k\}$ has been chosen to ensure such a property, and the expansion coefficients a_n are obtained explicitly interms of $f(y_k)$, $k = 0, 1, ..., N$, as described in [8]. Then, after substituting the expansion coefficients, the approximation given by Eq. (5) can be rearranged as [7]

$$f_N(y) = \sum_{k=0}^{N} f(y_k) h_k(y). \tag{18}$$

Equation (18) is simply an approximation to $f(y)$ in terms of the values of $f(y)$ at the collocation points y_k and Lagrangian interpolants $h_k(y)$ which are polynomials of degree at most N.

In the Chebyshev–Gauss–Lobatto collocation scheme, $\phi_n(y) = T_n(y)$ and the collocation points in the interval (-1, 1) are chosen to be the extrema

$$y_k = \cos\frac{\pi k}{N}, \quad k = 0, 1, ..., N \tag{19}$$

of the Nth–order Chebyshev polynomials $T_N(y)$. The Lagrangian interpolants are given by

$$h_k(y) = \frac{(1 - y^2) \, T_N'(y) \, (-1)^{k+1}}{c_k \, N^2 \, (y - y_k)}, \tag{20a}$$

or

$$h_k(y) = \frac{2}{N c_k} \sum_{n=0}^{N} \frac{1}{c_n} T_n(y_k) T_n(y) \qquad (20b)$$

where prime denotes differentiation, and

$$c_k = \begin{cases} 2, & \text{for } k = 0 \text{ or } N; \\ 1, & \text{for } 1 \leq k \leq N - 1. \end{cases} \qquad (21)$$

In the Legendre–Gauss–Lobatto collocation scheme $\phi_n(y) = L_n(y)$ and $y_0 = -1$, $y_N = 1$ and y_k ($k = 1, ..., N - 1$) are the zeros of $L'_N(y)$, where $L_n(y)$ is the Legendre polynomial of degree N. The Lagrangian interpolants are given by

$$h_k(y) = -\frac{(1 - y^2) L'_N(y)}{N(N + 1) L_N(y_k)(y - y_k)}. \qquad (22)$$

The properties of Chebyshev and Legendre polynomials are summarized in Appendices A and B, respectively.

2.4 Applications to Ordinary Differential Equations

To illustrate the difference between these methods, we begin by describing how to construct different spectral approximations to the following boundary value problem:

$$\mathcal{L}u = f \qquad (23)$$

on the interval $\Lambda = [-1, 1]$ with the homogeneous boundary conditions, $u(\pm 1) = 0$. Here $\mathcal{L}$ is a linear differential (spatial) operator.

The discrete approximations to the differential equation given by Eq. (23) can be written as

$$\mathcal{L}u_N = f, \qquad (24)$$

where the approximation u_N is defined in the form of a truncated series as

$$u_N = \sum_{n=0}^{N} \hat{u}_n \phi_n(y). \qquad (25)$$

A *Galerkin* approximation is then constructed as follows:

$$(\mathcal{L}u_N, \phi_n)_w = (f, \phi_n)_w, \quad n = 0, 1, ..., N, \qquad (26)$$

where

$$\phi_n(\pm 1) = 0. \qquad (27)$$

The expansion coefficients $\hat{u}_n$ are determined by rewriting Eq. (26) in the following form.

$$\int_{-1}^{1} (\mathcal{L}u_N - f) \, \phi_n(y) \, w(y) \, dy = 0, \quad n = 0, 1, ..., N. \tag{28}$$

This gives a system of $N + 1$ ordinary differential equations for the $N + 1$ unknown coefficients, $\hat{u}_n$, $n = 0, 1, ..., N$.

A *tau* approximation, on the other hand, is constructed as follows:

$$(\mathcal{L}u_N, P_n) = (f, P_n), \quad P_n \in \mathcal{P}_N(\Lambda), \tag{29}$$

where $\mathcal{P}_N(\Lambda)$ is the space of all polynomials of degree $\leq N$ on the interval Λ. As usual, u_N are defined as

$$u_N(y) = \sum_{n=0}^{N} \hat{u}_n P_n(y). \tag{30}$$

Substituting Eq. (30) into Eq. (24), multiplying both sides of the resultant equation by $P_m(y)$, and integrating the resultant equation from -1 to 1, we obtain $N - 1$ equations (since derivative operator is a lowering order operator, the second derivative operator maps the set of basis functions $P_n(y) : n = 0, ..., N$ into the set $P_n(y) : n = 0, ..., N - 1$). Now, we add the boundary conditions

$$u_N(\pm 1) = \sum_{n=0}^{N} \hat{u}_n P_n(\pm 1) = 0. \tag{31}$$

As before, we end up with $N + 1$ equations for the $N + 1$ unknown spectral coefficients, $\hat{u}_n, n = 0, 1, ..., N$.

A *collocation* approximation is constructed as follows:

$$\mathcal{L}u_N(y_k) = f(y_k), \tag{32}$$

where u_N are given by

$$u_N(y) = \sum_{k=0}^{N} u(y_k) h_k(y). \tag{33}$$

Here $u(y_k)$ is the value of u at the collocation point y_k.

Spectral collocation approximations lead to a linear system

$$\mathbf{LU} = \mathbf{F}, \tag{34}$$

where $\mathbf{U}$ and $\mathbf{F}$ are vectors consisting of the grid point values of u, f and any boundary data, that is,

$$\mathbf{U} = [u_0, u_1, \ldots, u_N]^T \quad \text{and} \quad \mathbf{F} = [f_0, f_1, \ldots, f_N]^T,$$

and $\mathbf{L}$ is a matrix (constructed as a tensor product matrix in two or more spatial dimensions). A similar linear system is obtained for Galerkin and tau approximations, but now $\mathbf{U}$ and $\mathbf{F}$ are vectors consisting of the expansion coefficients of u, f, that is,

$$\mathbf{U} = [\hat{u}_0, \hat{u}_1, \ldots, \hat{u}_N]^T, \qquad \mathbf{F} = [\hat{f}_0, \hat{f}_1, \ldots, \hat{f}_N]^T$$

and the boundary data, and $\mathbf{L}$ is the appropriate matrix in transform space. Hereafter, we restrict our discussions entirely to collocation methods, mainly because they offer the simplest treatment for both linear and non–linear equations with variable coefficients. Note that boundary conditions will modify $\mathbf{L}$, $\mathbf{U}$ and $\mathbf{F}$ slightly.

3. Derivatives in the Collocation Method

The next step, after approximating the function $f(y)$ by an Nth degree polynomial $f_N(y)$, is to express the derivatives of $f_N(y)$ in terms of $f(y)$ at the collocation points y_i. This means that in a spectral collocation method the derivative at a point is approximated by means of a polynomial whose degree is equal to the total number of collocation points. So that the spectral approximation has a global character, contrary to finite-difference (or finite element) methods in which the approximation has a local character.

Let us now consider the pth derivative. The pth derivative, being a linear operator, can be approximated as

$$\frac{d^p}{dy^p} f \cong \frac{d^p}{dy^p} f_N. \tag{35a}$$

Therefore,

$$\frac{d^p}{dy^p} f_N \big|_{y=y_i} = \sum_{j=0}^{N} \frac{d^p h_j(y)}{dy^p} \big|_{y=y_i} f(y_j) = \sum_{j=0}^{N} D_{i,j}^{(p)} f(y_j) \tag{35b}$$

where

$$D_{i,j}^{(p)} = \frac{d^p h_j(y)}{dy^p} \big|_{y=y_i} \tag{36}$$

is the pth order differentiation matrix.

The first–order differentiation matrix matrix $D_{s,p}^{(1)}$ transforms a vector of data at the collocation points into approximate derivatives at those points:

$$D_{s,p}^{(1)} \begin{bmatrix} f_0 \\ \vdots \\ f_N \end{bmatrix} \cong \begin{bmatrix} f_0' \\ \vdots \\ f_N' \end{bmatrix}. \tag{37}$$

3.1 Chebyshev Approximation

3.1.1 Chebyshev Differentiation Matrices

To obtain Chebyshev differentiation matrices we differentiate the relation (20a). Thus, the first-order Chebyshev differentiation matrix $D_{s,p}^{(1)}$ has the entries

$$D_{0,0}^{(1)} = -D_{N,N}^{(1)} = \frac{2N^2 + 1}{6} \tag{38a}$$

$$D_{i,i}^{(1)} = -\frac{y_i}{2(1 - y_i^2)} \quad \text{for} \quad 1 \le i \le N - 1 \tag{38b}$$

$$D_{i,j}^{(1)} = \frac{c_i}{c_j} \frac{(-1)^{i+j}}{(y_i - y_j)} \quad \text{for} \quad 0 \le i, j \le N, i \ne j \tag{38c}$$

where

$$c_i = \begin{cases} 2, & \text{for i=0 or N;} \\ 1, & \text{for } 1 \le i \le N - 1. \end{cases}$$

The second-order spectral differentiation matrix $D^{(2)}$ is not identical to the square of $D^{(1)}$, *i.e.*, $D^{(2)} \ne [D^{(1)}][D^{(1)}]$. However, for most values of N, the matrix product $[D^{(1)}][D^{(1)}]$ would serve as a good discrete second–order differentiation operator. Entries of $D_{s,p}^{(2)}$ are given in [8] as

$$D_{i,j}^{(2)} = \frac{(-1)^{i+j}}{c_j} \frac{y_i^2 + y_i y_j - 2}{(1 - y_i^2)(y_i - y_j)^2} \quad \text{for} \quad 1 \le i, j \le N - 1, i \ne j \tag{39a}$$

$$D_{i,i}^{(2)} = \frac{(N^2 - 1)(1 - y_i^2) + 3}{3(1 - y_i^2)^2} \quad \text{for} \quad 1 \le i \le N - 1 \tag{39b}$$

$$D_{0,j}^{(2)} = \frac{2}{3} \frac{(-1)^j}{c_j} \frac{(2N^2 + 1)(1 - y_j) - 6}{(1 - y_j)^2} \quad \text{for} \quad 1 \le j \le N - 1 \tag{39c}$$

$$D_{N,j}^{(2)} = \frac{2}{3} \frac{(-1)^{N+j}}{c_j} \frac{(2N^2 + 1)(1 + y_j) - 6}{(1 + y_j)^2} \quad \text{for} \quad 1 \le j \le N - 1 \tag{39d}$$

$$D_{0,0}^{(2)} = D_{N,N}^{(2)} = \frac{N^4 - 1}{15} \tag{39e}$$

3.1.2 Chebyshev Differentiation by FFT

A different way to obtain an expression for the derivative of $f_N(y)$ is to differentiate the relation (20b) to get

$$f_N(y) = \sum_{n=0}^{N} a_n T_n(y), \tag{40}$$

where [7]

$$a_n = \frac{2}{N} \frac{1}{c_n} \sum_{j=0}^{N} \frac{f(y_j)T_n(y_j)}{c_j} = \frac{2}{N} \frac{1}{c_n} \sum_{j=0}^{N} \frac{f(y_j)}{c_j} \cos \frac{\pi j n}{N}. \qquad (41)$$

3.2 Legendre Collocation Derivative

The first-order differentiation matrix has entries

$$D_{i,j}^{(1)} = \frac{L_N(y_i)}{L_N(y_j)} \frac{1}{(y_i - y_j)}, \qquad i \neq j \qquad (42)$$

$$D_{i,i}^{(1)} = 0, \qquad i \neq 0, N \qquad (43)$$

$$D_{N,N}^{(1)} = \frac{1}{4} N(N + 1) = -D_{0,0}^{(1)} \qquad (44)$$

The difference between the Chebyshev and Legendre methods is evident here. The matrix $D_{s,p}^{(1)}$ for Legendre polynomials is nearly antisymmetric, in contrast to the Chebyshev matrix given in Eqs. (38).

The second-order differentiation matrix $D_{s,p}^{(2)}$ can be obtained by the matrix product $[D^{(1)}][D^{(1)}]$. The entries of $D_{s,p}^{(2)}$ have been given by Gottlieb, *et al.* [5] as

$$D_{i,j}^{(2)} = -2 \frac{L_N(y_i)}{L_N(y_j)} \frac{1}{(y_i - y_j)^2}, \qquad (1 \leq i, j \leq N - 1, \ i \neq j) \qquad (45a)$$

$$D_{i,i}^{(2)} = -\frac{1}{3} \frac{N}{(1 - y_i^2)}, \qquad (1 \leq i \leq N - 1) \qquad (45b)$$

These show that $D_{s,p}^{(2)} = \Lambda S \Lambda^{-1}$, where Λ is a diagonal and S is a symmetric matrix.

4. Applications

In this section, first a survey of the literature on the applications of spectral methods to heat transfer problems is given and then a representative example is presented.

4.1 Literature Survey

Spectral methods have been used to study, among others, fundamental instabilities which trigger transition from laminar to turbulent flow. Lee *et al.* [11] investigated the stability of fluids in rectangular enclosures with arbitrary aspect ratios by using both the Chebyshev pseudospectral and Galerkin methods. The linear stability of mixed convection in an annulus at large aspect ratio of 100 was studied by Yao and Rogers [12] using pseudospectral Chebyshev method. Also recently, Rogers and Yao [13] worked on natural convection in a heated vertical concentric annulus by using spectral Chebyshev collocation technique to discretize the equations for the basic-state and disturbance. Desrayaud [14] and his co-workers [15–17] have used spectral tau method to study the effects of radiation on the stability of natural convection of an absorbing–emitting, non–gray and non–scattering fluid contained in a vertical slot having isothermal side walls at different temperatures. Recently, Zenouzi [18] and Zenouzi and Yener [19, 20] implemented Legendre–collocation scheme to study the base flow and the stability of natural convection of a radiating viscous fluid in a vertical narrow slot having isothermal side walls of different temperatures. In the area of heat transfer in porous media, Rao et al. [21] analyzed natural convection in a horizontal porous annulus heated from the inner surface with the Galerkin spectral method. Rao et al. [22] also used a 3D Galerkin scheme expanding the temperature and the potential vector fields into Fourier series to study the natural convection in a fluid saturated porous annulus heated from the inner surface. Caltagirone and Bories [23] used a three-dimensional numerical model based upon a Galerkin spectral method to study the stability criteria of natural convective flow in an inclined porous layer. More recently, Charrier-Mojtabi et al. [24] investigated two-dimensional free convection flows in an annular porous layer numerically using both Fourier-Galerkin and collocation Chebyshev methods. Their numerical results indicate that the collocation-Chebyshev method gives a better accuracy especially for the description of the boundary layers developed near the inner and outer cylinders. In the area of cooling of electronic devices, Ghaddar et al. [25] investigated, by direct numerical simulation using the spectral element method, the stability and self-sustained oscillations of incompressible moderate–Reynolds–number flow in periodically grooved channels. Amon and Mikic [26] used the spectral element technique to solve the unsteady equations governing flow and heat transfer in slotted channels. Karniadakis [27] used a general-purpose spectral element code for simulation of unsteady Navier–Stokes and energy equations for a two-dimensional laminar flow past a circular cylinder. Hatziavramidis and Ku [28] studied laminar flow heat transfer problems associated with pipeline transport of oil using Chebyshev pseudospectral method.

From these examples, it is clear that spectral methods are well suited to heat transfer problems where hydrodynamic and thermal instabilities are important to the accurate prediction of heat transfer rates. It is in these cases that traditional techniques and models are inadequate. This is a consequence of the difficulty of creating a model of an instability and the fact that accurate prediction of hydrodynamic and thermal instabilities requires extremely good spatial resolution.

4.2 Example

We present here an example that involves mixed convection in a horizontal two-dimensional channel. The flow is assumed periodic in the streamwise x–direction with periodicity L_x. The non-dimensional distance between the parallel plates is 2.0 ($-1 \leq y \leq 1$). Zero velocity is imposed at the two horizontal plates. A non-dimensional temperature of $\theta = 0$ is imposed at the upper surface while a spatially periodic boundary condition $\theta_l(x)$ is imposed at the lower surface. The non-dimensional governing equations are:

$$\frac{\partial \mathbf{U}^*}{\partial t} + \omega^* \times \mathbf{U}^* = -\nabla p^* + \frac{1}{Re}\nabla^2 \mathbf{U}^* + \frac{Ra}{Re^2 Pr}\theta^* \hat{\mathbf{j}} \qquad (46a)$$

$$\nabla \cdot \mathbf{U}^* = 0 \qquad (46b)$$

$$\frac{\partial \theta^*}{\partial t} + \mathbf{U}^* \cdot \nabla \theta^* = \frac{1}{RePr}\nabla^2 \theta^* \qquad (46c)$$

The important non-dimensional parameters governing the flow are Rayleigh number, Reynolds number, and Prandtl number and are defined as follows:

$$Ra = \frac{H^3 \beta g \Delta T}{\alpha \nu}, \quad Re = \frac{H\bar{U}}{\nu} \quad \text{and} \quad Pr = \frac{\nu}{\alpha}$$

The lengths have been non-dimensionalized with respect to the channel half height, H; the velocity to the mean velocity, $\bar{U}$; and the temperature to the maximum temperature variation, $\theta_{max} - \theta_{min}$. In the following equations, the asterisks are dropped and all variables are assumed to be non-dimensional.

The Chebyshev-Tau method that is employed here is essentially the same as the technique used by Kim et al. [29] for simulations of turbulent channel flows. For this application we have added the energy equation and the buoyancy term to the momentum equation. Only two-dimensional flows are considered. In this scheme, the pressure is eliminated by taking the curl of the Navier-Stokes equation. The x– and z–direction components of the resulting vorticity equation are combined to form a fourth order equation for the v–velocity, which is solved with the remaining y–component of the vorticity equation and the energy equation. The governing equations can be written in the following form:

$$\frac{\partial u}{\partial t} = -\frac{\partial p}{\partial x} + h_1 + \frac{1}{Re}\nabla^2 u \qquad (47a)$$

$$\frac{\partial v}{\partial t} = -\frac{\partial p}{\partial y} + h_2 + \frac{1}{Re}\nabla^2 v + \frac{Ra}{Re^2 Pr}\theta \qquad (47b)$$

$$\frac{\partial w}{\partial t} = -\frac{\partial p}{\partial z} + h_3 + \frac{1}{Re}\nabla^2 w \qquad (47c)$$

$$\frac{\partial u}{\partial x} + \frac{\partial v}{\partial y} + \frac{\partial w}{\partial z} = 0 \qquad (47d)$$

$$\frac{\partial \theta}{\partial t} = -(u\frac{\partial \theta}{\partial x} + v\frac{\partial \theta}{\partial y} + w\frac{\partial \theta}{\partial z}) + \frac{1}{RePr}\nabla^2\theta \qquad (47e)$$

where h_i includes the convective terms, defined as

$$h_i = \epsilon_{ijk}u_j\omega_k - \frac{1}{2}\frac{\partial}{\partial x_j}(u_j u_j) \qquad (48)$$

The symmetric part $-\frac{1}{2}\frac{\partial}{\partial x_j}(u_j u_j)$ will drop out of the equation later when we apply the differential operator to h_i. Here, we just write down h_i as:

$$h_1 = v\omega_z - w\omega_y \qquad (49a)$$

$$h_2 = w\omega_x - u\omega_z \qquad (49b)$$

$$h_3 = u\omega_y - v\omega_x \qquad (49c)$$

where ω_x, ω_y, ω_z are vorticity components in x–, y– and z–directions, respectively. They are defined as

$$\omega_x = \frac{\partial w}{\partial y} - \frac{\partial v}{\partial z} \qquad (50a)$$

$$\omega_y = \frac{\partial u}{\partial z} - \frac{\partial w}{\partial x} \qquad (50a)$$

$$\omega_z = \frac{\partial v}{\partial x} - \frac{\partial u}{\partial y} \qquad (50a)$$

The problem now has been reduced to the following set of equations:

$$\frac{\partial \Phi}{\partial t} = h_v + Re^{-1}\nabla^2\Phi \qquad (51a)$$

$$\nabla^2 v = \Phi \qquad (51b)$$

$$\frac{\partial G}{\partial t} = h_g + Re^{-1}\nabla^2 G \qquad (51c)$$

$$f + \frac{\partial v}{\partial y} = 0 \qquad (51d)$$

$$\frac{\partial \theta}{\partial t} = h_T + (RePr)^{-1}\nabla^2\theta \qquad (51e)$$

where

$$f = \frac{\partial u}{\partial x} + \frac{\partial w}{\partial z} \qquad (52a)$$

$$G = \frac{\partial u}{\partial z} - \frac{\partial w}{\partial x} \qquad (52b)$$

$$h_g = \frac{\partial H_1}{\partial z} - \frac{\partial H_3}{\partial x} \qquad (52c)$$

$$h_v = -\frac{\partial}{\partial y}\left(\frac{\partial H_1}{\partial x} + \frac{\partial H_3}{\partial z}\right) + \left(\frac{\partial^2}{\partial x^2} + \frac{\partial^2}{\partial z^2}\right)H_2 \qquad (52d)$$

$$h_T = -\left(u\frac{\partial \theta}{\partial x} + v\frac{\partial \theta}{\partial y} + w\frac{\partial \theta}{\partial z}\right) \qquad (52e)$$

and

$$H_1 = h_1 - \frac{\partial p}{\partial x} \qquad (52f)$$

$$H_2 = h_2 + \frac{Ra}{Re^2 Pr}\theta \qquad (52g)$$

$$H_3 = h_3 \qquad (52h)$$

with corresponding boundary conditions:

$$v = 0 \quad at \quad y = \pm 1 \qquad (53a)$$

$$\frac{\partial v}{\partial y} = 0 \quad at \quad y = \pm 1 \qquad (53b)$$

$$G = 0 \quad at \quad y = \pm 1 \qquad (53c)$$

$$\theta = \theta_1 \quad at \quad y = 1 \qquad (53d)$$

$$\theta = \theta_2 \quad at \quad y = -1 \qquad (53e)$$

and periodic boundary conditions in x– and z–directions. This system is now ready to be solved by a spectral method – Fourier series in streamwise and spanwise direction, and Chebyshev polynomial expansion in normal direction. The solutions have the form:

$$\mathbf{U} = \sum_{|j|<N_x} \sum_{|k|<N_z} \sum_{l=0}^{N_y} \hat{\hat{u}}(j,k,l,t)exp[2\pi i(\frac{jx}{N_x} + \frac{kz}{N_z})]T_l(y) \qquad (54a)$$

$$\theta = \sum_{|j|<N_x} \sum_{|k|<N_z} \sum_{l=0}^{N_y} \hat{\hat{\theta}}(j,k,l,t)exp[2\pi i(\frac{jx}{N_x} + \frac{kz}{N_z})]T_l(y) \qquad (54b)$$

where $T_l(y) = \cos(l\cos^{-1}y)$ is the Chebyshev polynomial of degree l.

4.2.1 Temporal Discretization

Crank-Nicolson and Adams-Bashforth methods are employed for the diffusive and convective terms, respectively. The equation for θ becomes

$$\frac{\theta^{n+1} - \theta^n}{\Delta t} = \frac{3}{2}h_T^n - \frac{1}{2}h_T^{n-1} + \frac{1}{RePr}\left(\frac{\nabla^2\theta^{n+1} + \nabla^2\theta^n}{2}\right) \tag{55a}$$

which also can be written as

$$\nabla^2\theta^{n+1} - \frac{2RePr}{\Delta t}\theta^{n+1} = -3RePrh_T^n + RePrh_T^{n-1} - \nabla^2\theta^n + \frac{2RePr}{\Delta t}\theta^n) \tag{55b}$$

$$\theta = \theta_1 \quad at \quad y = 1 \tag{55c}$$

$$\theta = \theta_2 \quad at \quad y = -1 \tag{55d}$$

Similarly, the equation for G becomes

$$\frac{G^{n+1} - G^n}{\Delta t} = \frac{3}{2}h_g^n - \frac{1}{2}h_g^{n-1} + \frac{1}{Re}\left(\frac{\nabla^2G^{n+1} + \nabla^2G^n}{2}\right) \tag{56a}$$

$$\nabla^2G^{n+1} - \frac{2Re}{\Delta t}G^{n+1} = -3Reh_T^n + Reh_T^{n-1} - \nabla^2G^n + \frac{2Re}{\Delta t}G^n) \tag{56b}$$

$$G = 0 \quad at \quad y = \pm 1 \tag{56c}$$

and equation for v becomes

$$\nabla^2\Phi_p^{n+1} - \frac{2Re}{\Delta t}\Phi_p^{n+1} = -3Reh_v^n + Reh_v^{n-1} - \nabla^2\Phi^n + \frac{2Re}{\Delta t}\Phi^n \tag{57a}$$

$$\Phi_p = 0 \quad at \quad y = \pm 1 \tag{57b}$$

$$\nabla^2v_p^{n+1} = \Phi_p^{n+1} \tag{57c}$$

$$v_p = 0 \quad at \quad y = \pm 1 \tag{57d}$$

$$\nabla^2\Phi_1^{n+1} - \frac{2Re}{\Delta t}\Phi_1^{n+1} = 0 \tag{58a}$$

$$\Phi_1 = 0 \quad at \quad y = 1 \tag{58b}$$

$$\Phi_1 = 1 \quad at \quad y = -1 \tag{58c}$$

$$\nabla^2v_1^{n+1} = \Phi_1^{n+1} \tag{58d}$$

$$v_1 = 0 \quad at \quad y = \pm 1 \tag{58e}$$

$$\nabla^2 \Phi_2^{n+1} - \frac{2Re}{\Delta t} \Phi_2^{n+1} = 0 \tag{59a}$$

$$\Phi_2 = 0 \quad at \quad y = 1 \tag{59b}$$

$$\Phi_2 = 1 \quad at \quad y = -1 \tag{59c}$$

$$\nabla^2 v_2^{n+1} = \Phi_2^{n+1} \tag{59d}$$

$$v_2 = 0 \quad at \quad y = \pm 1 \tag{59e}$$

The temporal discretization is second order accurate.

4.2.2 Spatial Discretization

Fourier series is used in streamwise and spanwise directions, and Chebyshev polynomial expansion is used in normal direction. The grid points in x–and z–direction are defined as

$$x_i = \frac{iL_x}{N_x} \quad i = 0, 1, ..., N_x - 1 \tag{60a}$$

$$z_k = \frac{kL_z}{N_z} \quad k = 0, 1, ...N_z - 1 \tag{60b}$$

The grid pojnts for y direction are defined as Chebyshev-Gauss-Lobatto points

$$y_j = cos\frac{\pi j}{N_y} \quad j = 0, 1, ..., N_y \tag{60c}$$

The system is solved in Fourier-Chebyshev space. We denote $\tilde{\theta}$ as the Fourier coefficient of θ, $\hat{\theta}$ as the Chebyshev coefficient of θ, and D^2 as the second derivative with respect to y. The equation of θ in Fourier-Chebyshev space then becomes

$$D^2 \hat{\tilde{\theta}}^{n+1} - (\alpha^2 + \beta^2)\hat{\tilde{\theta}}^{n+1} - \frac{2RePr}{\Delta t}\hat{\tilde{\theta}}^{n+1} = RHS_T^{n,n-1} \tag{61a}$$

where

$$RHS_T = -3RePr\hat{\tilde{h}}_T^{\ n} + RePr\hat{\tilde{h}}_T^{\ n-1} - D^2\hat{\tilde{\theta}}^n + (\alpha^2 + \beta^2)\hat{\tilde{\theta}}^n + \frac{2RePr}{\Delta t}\hat{\tilde{\theta}}^n \tag{61b}$$

and the boundary conditions are

$$\hat{\tilde{\theta}} = \hat{\tilde{\theta}}_1 \quad at \quad y = 1 \tag{61c}$$

$$\hat{\tilde{\theta}} = \hat{\tilde{\theta}}_2 \quad at \quad y = -1 \tag{61d}$$

Here α and β are wavenumbers in x-and z-direction, respectively. We define

$$\kappa^2 = \alpha^2 + \beta^2 \tag{62}$$

and rewrite Eq. (61a) as

$$D^2\hat{\tilde{\theta}}^{n+1} - (\kappa^2 + \frac{2RePr}{\Delta t})\hat{\tilde{\theta}}^{n+1} = RHS_T^{n,n-1} \tag{63}$$

which is a second-order equation in Fourier-Chebyshev space, and has to be solved with each α and β.

Using the same procedure as we used to θ equation, we have

$$D^2\hat{\tilde{G}}^{n+1} - (\kappa^2 + \frac{2Re}{\Delta t})\hat{\tilde{G}}^{n+1} = RHS_g^{n,n-1} \tag{64a}$$

$$\hat{\tilde{G}} = 0 \quad at \quad y = \pm 1 \tag{64b}$$

$$RHS_g = -3Re\hat{\tilde{h}}_g^{\,n} + Re\hat{\tilde{h}}_g^{\,n-1} - D^2\hat{\tilde{G}}^n + (\alpha^2 + \beta^2)\hat{\tilde{G}}^n + \frac{2Re}{\Delta t}\hat{\tilde{G}}^n \tag{64c}$$

and

$$D^2\hat{\tilde{\Phi}}_p^{\,n+1} - (\kappa^2 + \frac{2Re}{\Delta t})\hat{\tilde{\Phi}}_p^{\,n+1} = RHS_v^{n,n-1} \tag{65a}$$

$$RHS_v = -3Re\hat{\tilde{h}}_v^{\,n} + Re\hat{\tilde{h}}_v^{\,n-1} - D^2\hat{\tilde{\Phi}}^n + (\alpha^2 + \beta^2)\hat{\tilde{\phi}}^n + \frac{2Re}{\Delta t}\hat{\tilde{\Phi}}^n \tag{65b}$$

$$\hat{\tilde{\Phi}}_p = 0 \quad at \quad y = \pm 1 \tag{65c}$$

$$D^2\hat{\tilde{v}}_p^{\,n+1} - \kappa^2\hat{\tilde{v}}_p^{\,n+1} = \hat{\tilde{\Phi}}_p^{\,n+1} \tag{65d}$$

$$\hat{\tilde{v}}_p^{\,n+1} = 0 \quad at \quad y = \pm 1 \tag{65e}$$

$$D^2\hat{\tilde{\Phi}}_1^{\,n+1} - (\kappa^2 + \frac{2Re}{\Delta t})\hat{\tilde{\Phi}}_1^{\,n+1} = 0 \tag{66a}$$

$$\hat{\tilde{\Phi}}_1 = 0 \quad at \quad y = 1 \tag{66b}$$

$$\hat{\tilde{\Phi}}_1 = 1 \quad at \quad y = -1 \tag{66c}$$

$$D^2\hat{\tilde{v}}_1^{\,n+1} - \kappa^2\hat{\tilde{v}}_1^{\,n+1} = \hat{\tilde{\Phi}}_1^{\,n+1} \tag{66d}$$

$$\hat{\tilde{v}}_1^{\,n+1} = 0 \quad at \quad y = \pm 1 \tag{66e}$$

$$D^2 \hat{\bar{\Phi}}_2^{\,n+1} - (\kappa^2 + \frac{2Re}{\Delta t}) \hat{\bar{\Phi}}_2^{\,n+1} = 0 \qquad (67a)$$

$$\hat{\bar{\Phi}}_2 = 1 \qquad at \quad y = 1 \qquad (67b)$$

$$\hat{\bar{\Phi}}_2 = 0 \qquad at \quad y = -1 \qquad (67c)$$

$$D^2 \hat{\bar{v}}_2^{\,n+1} - \kappa^2 \hat{\bar{v}}_2^{\,n+1} = \hat{\bar{\Phi}}_2^{\,n+1} \qquad (67d)$$

$$\hat{\bar{v}}_2^{\,n+1} = 0 \qquad at \quad y = \pm 1 \qquad (67e)$$

The transformation between Fourier, Chebyshev and physical space is evaluated by a Real FFT (Fast Fourier Transform). The transforms related to Chebyshev space are also able to use FFT because of the Gauss-Lobatto points chosen in y–direction, which make the discrete Chebyshev expansion become a simple cosine series.

The FFT used in this work is a real transform. Because all the functions are real and a cosine transform for Chebychev transform is desired, the direct use of the complex FFT is needlessly expensive. The complex coefficients of a function in Fourier space are therefore split into two real sets of data and calculated separately.

The scheme derived here is for three–dimensional flow. However, the solutions presented are two–dimensional, so we take $N_z = 1$.

We first consider a special case of this flow, with $Re = 0$. Figure 2 shows the streamlines and isotherms for this flow with $Ra = 800$ and $Pr = 1$. Note that the flow appears similar to Rayleigh-Benard Convection. There is no critical Rayleigh number since the temperature varies in the horizontal direction so that there will always be some fluid motion.

Spectral methods are particularly valuable in simulating flow instabilities, since these flows tend to be more complex and require greater accuracy. This two-dimensional mixed convection flow has been studied by Zhang and Tangborn [30]. Instabilities that depend upon the values of Ra, Re and the channel length L_x have been observed. One example of the resulting unsteady flow during one oscillation cycle is shown in Figures 3 and 4, for the case $Ra = 60,000$, $Re = 20$ and $Pr = 1$. The streamlines show that a vortex has formed downstream of the hot section of the wall, which periodically grows and sheds from the wall. The isotherms show that this fluid motion creates plumes of hot fluid that rise periodically from the lower wall and are carried down stream. This type of flow shows the potential for enhanced heat transfer and mixing in applications such as electronic cooling or materials processing. This particular example is representative of the type of problem for which spectral methods are especially well suited: Flow at moderate Reynolds number in a simple geometry in which flow instabilities occur that result in complicated flow patterns.

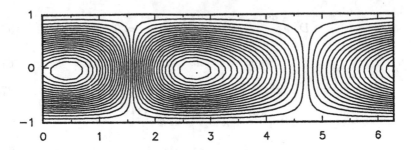

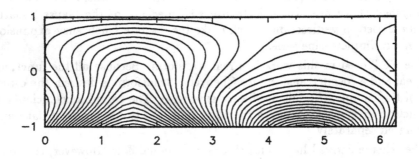

Fig. 2. Streamlines and isotherms; $Ra = 800$, $Re = 0$ and $Pr = 1$.

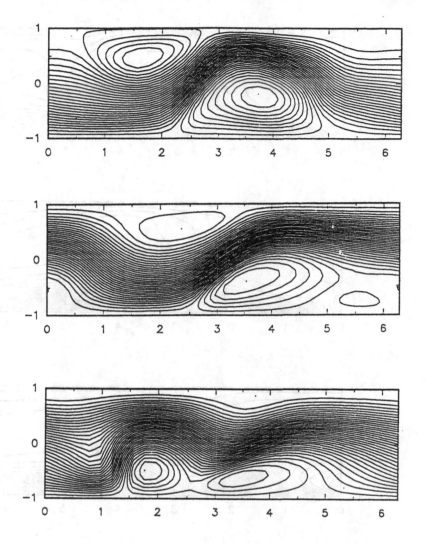

Fig. 3. Streamlines during one cycle of oscillation. $Ra = 60,000$, $Re = 20$ and $Pr = 1$.

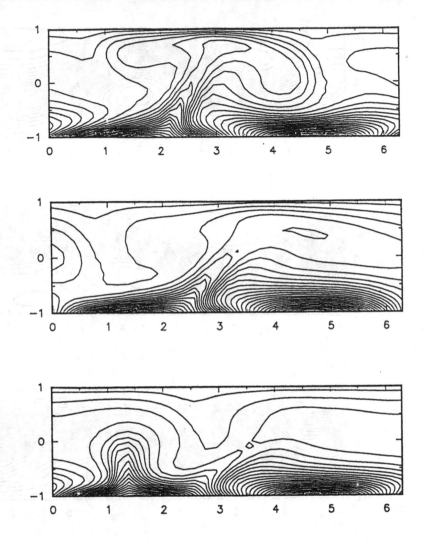

Fig. 4. Isotherms during one cycle of oscillation. $Ra = 60,000$, $Re = 20$ and $Pr = 1$.

References

1. Patera, A.T.: Advances and future directions of research on spectral mehtods. In: Computational mechanics–advances and trends, A.K. Noor (ed.), ASME/AMD–Vol. 75, 411–427 (1987)

2. Fletcher, C.A.J.: Computational Galerkin methods. Springer–Verlag, New York (1984)

3. Finlayson, B.A.: The method of weighted residuals and variational principles. Academic Press, New York (1972)

4. Canuto, C., Hussaini, M.Y., Quarteroni, A., and Zang, T.A.: Spectral methods in fluid dynamics. Springer–Verlag, New York (1988)

5. Gottlieb, D., Hussaini, M.Y., and Orszag, S.A.: Theory and application of spectral methods. In: Spectral methods for partial differential equations, R.G. Voigt, D. Gottlieb, M.Y. Hussaini (eds.), SIAM-CBMS, Philadelphia, 1–54 (1984)

6. Gottlieb, D., and Orszag, S.A.: Numerical analysis of spectral methods: Theory and applications. SIAM-CBMS, Philadelphia (1977)

7. Gottlieb, D., and Turkel, E.: Topics in spectral methods for time dependent problems. In: Numerical methods in fluid dynamics, F. Brezzi (ed.), Springer–Verlag, New York, 115–155 (1985)

8. Peyret, R.: Introduction to specteral methods. In: von Karman Inst. Lect. Series 1986–04, Rhode–Saint Genese, Belgium (1986)

9. Bourke, W., McAvaney, B., Puri, K., and Thurling, R.: Global modelling of atmospheric flow by spectral methods. In: Methods in computational physics, Chang (ed.), Academic Press, New York, 17, 267–324 (1977)

10. Solomonoff, A., and Turkel, E.: Global collocation methods for approximation and the solution of partial differential equations. ICASE Rep. No. 86–60, NASA Langely Research Center, Hampton, VA (1986)

11. Lee, N.Y., Schultz, W.W., and Boyd, J.P.: Stability of fluid in a rectangular enclosure by spectral method. Int. J. Heat Mass Transfer, 32, 513–520 (1989)

12. Yao, L.S., and Rogers, B.B.: Mixed convection in an annulus of Large Aspect Ratio. J. of Heat Transfer, 111, 683–689 (1989)

13. Rogers, B.B., and Yao, L.S.: Natural convection in a heated annulus. Int. J. Heat Mass Transfer, 36, 35–47 (1993)

14. Desrayaud, G.: Analyse de stabilité linéaire dans un milieu semi-transparent. Détermination expérimentale des limites de stabilité dans un milieu transparent. These de Doctorat D'etat, Université Pierre et Marie Curie (Paris 6) (1987),

15. Desrayaud, G., and Lauriat G.: Natural convection of radiating fluid in a vertical layer. J. of Heat Transfer, 107, 710–712 (1985)

16 Desrayaud, G., and Lauriat G.: On the stability of natural convection of a radiating fluid in a vertical slot. Commun. Heat Mass Transfer, 11, 439–450 (1984)

17. Desrayaud, G., and Lauriat G.: Radiative influence on the stability of fluids enclosed in vertical cavities. Int. J. Heat Mass Transfer, 31, 1035–1048 (1988)

18. Zenouzi, M.: Simultaneous radiation and natural convection in vertical slots. Ph.D. Thesis, Mechanical Engineering Department, Northeastern University, Boston, MA (1990)

19. Zenouzi, M., and Yener, Y.: Simultaneous radiation and natural convection in vertical slots. In: Developments in Radiative Heat Transfer, S.T. Thynell (ed.), ASME/HTD–Vol. 203, 179–186 (1992)

20. Zenouzi, M., and Yener, Y.: Thermal stability of a radiation fluid in a vertical narrow slots. In: Stability of Convective Flows, P.G. Simpkins (ed.), ASME/HTD–Vol. 219, 1–7 (1992)

21. Rao, Y.F., Fukuda, K., and Hasegawa, S.: Steady and transient analysis of natural convection in a horizontal porous annulus with Galerkin method. J. of Heat Transfer, 109, 919–927 (1987)

22. Rao, Y.F., Fukuda, K., and Hasegawa, S.: A numerical study of three-dimensional natural convection in a horizontal porous annulus with Galerkin method. Int. J. Heat Mass Transfer, 31, 695–707 (1988)

23. Caltagirone, J.P., and Bories, S.: Solution and stability criteria of natural convective flow in an inclined porous layer. J. of Fluid Mech., 155, 267–287 (1985)

24. Charrier–Mojtabi, M.C., Mojtabi, A., Azaiez, M., and Labrosse, G.: Numerical and experimental study of multicellular free convection flows in an annular porous layer. Int. J. Heat Mass Transfer, 34, 3061–307 (1991)

25. Ghaddar, N.K., Korczak, K.Z., Mikic, B.B., and Patera, A.T.: Numerical investigation of incompressible flow in grooved channels. J. of Fluid Mech., 163, 99–127 (1986)

26. Amon, C.A., and Mikic, B.B.: Spectral element simulations of unsteady forced convective heat transfer: Application to compact heat exchanger geometries. Numerical Heat Transfer, Part A, 19, 1–19 (1991)

27. Karniadakis, G.E.: Numerical simulation of forced convection heat transfer from a cylinder in crossflow. Int. J. Heat Mass Transfer, 31, 107–118 (1988)

28. Hatziavramidis, D., and Ku, H.C.: Pseudospectral Solutions of Laminar Heat Transfer Problems in Pipelines. J. of Comp. Physics, 52, 414–424 (1983)

29. Kim, J., Moin P., and Moser, R.: Turbulence statistics in fully developed channel flow at low reynolds number. J. of Fluid Mech., 177, 133–156 (1987)

30. Zhang, S.Q., and Tangborn, A.V.: Flow regimes in two-dimensional mixed convection with spatially periodic lower wall heating. Submitted to Physics of Fluids (1993)

Appendices

A. Properties of Chebyshev Polynomials

The Chebyshev polynomial of degree n is defined by

$$T_n(y) = \cos(n\cos^{-1}y), \quad -1 \le y \le 1.$$

They can be generated from the recurrence relation

$$T_0(y) = 1, \quad T_1(y) = y$$

$$T_{n+1} = 2\,y\,T_n(y) - T_{n-1}(y), \quad n \ge 1$$

Thus,

$$T_2(y) = 2y^2 - 1, \quad T_3(y) = 4y^3 - 3y, \quad T_4(y) = 8y^4 - 8y^2 + 1$$

Also,

$$T_n(-y) = (-1)^n\,T_n(y), \quad T_n(\pm 1) = (\pm 1)^n$$

Chebyshev polynomials form a complete orthogonal set with respect to the weight function o $w(y) = (1 - y^2)^{-1/2}$, that is,

$$\int_{-1}^{1} T_n(y)T_m(y)(1 - y^2)^{-1/2}dy = \frac{\pi}{2}c_n\delta_{nm},$$

If y_k $(k = 1, \ldots, m)$ are the m zeros of $T_m(y)$, the discrete analog of the above orthogonality relation is given by

$$\sum_{k=0}^{m} \frac{1}{c_k}T_i(y_k)T_j(y_k) = \frac{1}{2}mc_n\delta_{mn},$$

where

$$c_i = \begin{cases} 2, & \text{for } i=0 \text{ or } N; \\ 1, & \text{for } 1 \le i \le N-1. \end{cases}$$

For indefinite integration of Chebyshev polynomials we have

$$\int T_n(y)dy = \begin{cases} \frac{T_{n+1}(y)}{2(n+1)} - \frac{T_{n-1}(y)}{2(n-1)} & (n \ge 2), \\ \frac{1}{4}\left[T_0(y) + T_2(y)\right] & (n = 1), \\ T_1(y) & (n = 0). \end{cases}$$

A. Properties of Legendre Polynomials

The Legendre polynomials $\{L_n(y), \ n = 0, 1, \ldots\}$ are the eigenfunctions of the following singular Strum-Liouville problem:

$$\frac{d}{dy}\left[(1 - y^2)\frac{d}{dy}L_n(y)\right] + n(n + 1)L_n(y) = 0,$$

with $L_n(1) = 1$. They are orthogonal with respect to a weight of unity, that is,

$$\int_{-1}^{1} L_n(y)L_m(y)dy = \frac{2}{2n + 1}\delta_{nm},$$

and satisfy the following three-term recurrence relation,

$$L_0(y) = 1, \qquad L_1(y) = y$$

$$(n + 1)L_{n+1}(y) = (2n + 1)yL_n(y) - nL_{n-1}(y).$$

Also,

$$L_n(x) = \frac{1}{2^n n!}\sum_{k=0}^{m}(-1)^k \frac{n!}{k!(n - k)!}\frac{(2n - 2k)!}{(n - 2k)!}x^{n-2k}$$

where $m = n/2$ if n is even and $m = (n - 1)/2$ if n is odd. It is sometimes useful to express $L_n(x)$ by Rodrigues' formula

$$L_n(y) = \frac{1}{2^n n!}\frac{d^n}{dx^n}(x^2 - 1)^n, \qquad n = 0, 1, 2, \ldots.$$

We also have the relations

$$(2n + 1)L_n(y) = L'_{n+1}(y) - L'_{n-1}(y)$$

$$(1 - y^2)L'_n(y) = nL_{n-1}(y) - nyL_n(y)$$

$$L_n(\pm 1) = (\pm 1)^n, \quad L'_n(\pm 1) = (\pm 1)^{n+1}\frac{1}{2}n(n + 1)$$

$$\left[\frac{d^m}{dy^m}L_n(y)\right]_{y=1} = \frac{(n + m)!}{2^m m!(n - m)!}$$

$$\int_0^1 L_n(y)dy = \begin{cases} 1 & (n = 0), \\ 0 & (n = 2, 4, 6, \ldots), \\ (-1)^{(n-1)}\frac{1}{n(n+1)}\frac{(1\cdot3\cdot5\ldots n)^2}{n!} & (n = 1, 3, 5, \ldots) \end{cases}$$

$$\int_y^1 L_n(y)dy = \frac{1}{2n + 1}[L_{n-1}(y) - L_{n+1}(y)] \qquad (n = 1, 2, \ldots)$$

PART FOUR

Workshops

An important part of the NATO week was the organisation of formal workshops to stimulate discussion. Two workshops were formed, one to discuss mathematical modelling in fluid flow and heat transfer, and the other to discuss the possible impact of computer algebra packages on engineering mathematics.

The delegates divided approximately fifty-fifty into the two workshops and in both very lively discussions took place. Each workshop had two three-hour sessions led by an elected chairman and a reporter. The results of these two sessions are presented here.

Computer Algebra Systems in Mathematical Education

F. Simons
Technical University of Eindhoven, Eindhoven, The Netherlands

The following is a short report on the discussions that were held in two sessions on the use of computer algebra systems in mathematical education. Many people participated, resulting in lively discussions in which many different opinions turned up. We decided not to try to formulate all these, sometimes conflicting, points of view but to look for the common opinions.

We discussed only about the general undergraduate mathematical education at university level. The discussion did not specifically consider computer algebra packages but more generally mathematical software that could be used when teaching undergraduate mathematics, ranging from special didactical software for explaining concepts and techniques to advanced technical software such as computer algebra packages and professional numerical and graphical software.

We all agreed that for many reasons the use of these software should be strongly stimulated but at the same time we realised that quite a lot of problems arise when one wants to introduce the use of computers in the classroom.

First we have the technical problem. The mathematics courses will be taught as a laboratory course. This implies that the classroom need to be provided with computers, and in the lecture room good projection facilities must be available. At the moment not many universities have these facilities. Hence the university administrators must be convinced that it is worth doing so. Also the licences for using certain packages have to be arranged.

In the discussion it became clear that mathematical software can be used in many different ways, ranging from

- a traditional course with educational software used to explain concepts and techniques, to
- a course revised to reflect the computer revolution of recent years, making full use of mathematical software systems.

The choice of what will be done with the software depends on the local circumstances, the average level of students and the goals of the course. It was also remarked that computer algebra has to be introduced gradually and not at once in one step. A start might be to add a computer algebra practicum to a traditional course to show that many of the operations also can be performed on a computer.

Introducing mathematical software rises many didactical problems. The accent in the courses will automatically be more on concepts and less on hand calculations. Computer aided teaching can play a role here. It can be used for explaining the concepts and enhancing the understanding. When computer algebra is used, the question arises what can be subtracted (drillwork on techniques!) and what should or could be added to the course.

Analytical and numerical techniques will be treated simultaneously if numerical software is used. A tuning with the courses in numerical analysis has to take place, e.g. leaving the elementary topics to the basic courses and the more advanced topics to the special course on numerical mathematics. A similar remark can be made for the programming courses. It might be an advantage to devote the programming course to programming in a computer algebra package instead of a computer language. At some places this already happens.

Mathematical Modelling in Fluid Flow

A.W. Bush[1] and R. Mattheys[2]

[1]*Reporter*, University of Teesside, UK
[2]*Chairman*, British Gas plc, UK

1 Introduction

A group of about twelve members of the NATO-ARW chose to attend the mathematical modelling (MM) in fluid flow sessions. The sessions consisted of the following:

- *A presentation on the industrial requirements of fluid flow courses*
- *A presentation on experiences of teaching Computational Fluid Dynamics (CFD)*
- *Discussion sessions on the role of MM in engineering courses.*
- *A software demonstration of a CFD code[3]*

2 Teaching Fluid Dynamics and MM

The discussion on modelling was extremely lively. Strong and often opposing views were expressed early in the discussions. Some degree of consensus was reached by the end of the discussion time. This is described in the conclusion section. The early debate ranged over the following two issues.

2.1 Teaching Fluid Dynamics

There were different and rather opposite views as follows:

- *W1*: The available exact solutions of the governing equations should be carefully derived so that students are well trained in the analytical approach.
- *W2*: Few exact solutions exist and those that do are special so the solution techniques cannot be generalised. It is better not to emphasise the derivation of the exact solutions. Instead state the form of the solutions and spend time analysing their meaning.

2.2 Teaching Mathematical Modelling

There were various views on the subject matter. They can be stated as follows:

- Modelling should be taught using problems which require no new mathematics
- Lecturers responsible for modelling courses should not be seen as experts. If they are then students are inhibited and will not take part in the modelling activity. Thus modelling cannot be integrated into another course which

---

[3]*The presentation on CFD appears in Chp.12.*

> involves taught material.
- Modelling should motivate the need for new material which the lecturer then teaches.
- Student led modelling takes too long to achieve the correct formulation of a problem.

Most of the group recognised that a difficulty associated with providing a carefully structured course which is lecturer led is that students found it difficult to deal with material outside the range of topics in the course. On the other hand student led modelling courses help the students to learn to think for themselves.

Some felt that a historical account of the development of a subject such as fluid mechanics would show the students that the final elegant mathematical structure which is now available to them evolved out of various incomplete and sometimes erroneous theories. The historical approach would encourage their own imperfect attempts at modelling processes.

Some of the group felt that a compromise approach was best where a short time is devoted to student led modelling so that the philosophy and methodology can be appreciated. This is then followed by lecturer led case studies which emphasise the modelling aspects but which are treated at a sufficient rate to enable a reasonable coverage of the discipline in the time available.

Some of the group felt that students are better placed to undertake modelling after having acquired skills in exact and numerical solution procedures.

3 Computers and Software

The group agreed on the need for computers and relevant software to be available for the students to explore various fluid flow and heat transfer processes. Concern was expressed over the possibility of students accepting these solutions without critically assessing their worth. It was vital that the following features were emphasised:
- has the solution converged?
- is the solution grid independent?
- have the boundary conditions been properly imposed?
- does the scheme use techniques which aid convergence but may reduce accuracy e.g. the "upwinding" procedure?

A view was expressed that some full viscous CFD solutions were unnecessarily expensive both from the point of view of the cost of the package and the computer processing time. Techniques which involved some analysis were sometimes cheaper, quicker and more accurate. The "panel method" was particularly mentioned in this regard.

4 Conclusions

- Modelling is an essential factor for Engineering Education. The ability for an engineer to take previously unseen systems and explain them in terms of well understood phenomena is an important element of both analysis and design.

- Modelling training must be given very early in any course and must be reinforced by continuing activity during that course. The engineer needs the breadth, skills and confidence which modelling provides in order to carry out his future professional activities.

- Modelling skills for engineering will develop along a path similar to the engineering course itself from case studies to project work and finally to design work itself.

- Industrial Placement such as embodied within the co-op scheme is a very important factor in building maturity, understanding and confidence with students in general and must be viewed as a key element in engineering training.

- Mathematics may frequently be viewed by engineers as a dry subject which requires *"passing"* rather than as a key element of his professional tool kit. Mathematics teaching for engineers must be carried out by persons with empathy for the application subject in order to reinforce the subjects' importance and relevance.

- Resourcing and particularly computing is extremely important. Mathematical models of practical importance frequently require sophisticated, powerful hardware systems to run them on. The correct choice of software tools and environment is also critical in being able to provide the student with user friendly and efficient systems capable of reinforcing their confidence and ease with the tools.

- For students to gain the maximum amount of benefit from a modelling course, modelling should be student led. The skills acquired would then be based upon experience and practise and not on formal teaching and would increase confidence and competence with their use.

- Modelling needs to be able to respond to technology changes and to educational changes. This extends down to primary and secondary education sectors where technological changes are progressing at a fast rate. University courses must reflect these changes and build upon the opportunities presented. An example of this is the exposure students of all ages have had to computer games; models based upon these games or using similar interfaces are readily understood and accepted by even very young students.

- *"Teachers need to be taught"* - the methodology of modelling must be placed in a correct framework for both teachers and taught; ongoing training to meet changing needs or methods must be an integral part of a course philosophy.

Summary

- This particular strand of the workshop was unable to design a specific course for mathematical fluid flow modelling in engineering due to the differing needs of various application areas.

- There was much discussion during the group meetings on the nature and philosophy of modelling itself which did not meet a full consensus resolution. A general feeling however was that modelling should be taught as a general skill with applications arising from a number of non-specific technical areas. The experiences so gained could then be reinforced by reference to the course at the appropriate times. The general modelling skills were felt to be very important in ensuring a broad based level of expertise which could, at need, be mapped onto many differing areas of activity.

- Finally, there is a definite requirement for future activity to draft up specific course contents and methods. This should involve students as well as teachers.

Author and Subject Index

NATO ASI Series F

NATO ASI Series F

NATO ASI Series F

Including Special Programmes on Sensory Systems for Robotic Control (ROB) and on Advanced Educational Technology (AET)

NATO ASI Series F

NATO ASI Series F

NATO ASI Series F

Including Special Programmes on Sensory Systems for Robotic Control (ROB) and on Advanced Educational Technology (AET)

Springer-Verlag
and the Environment

We at Springer-Verlag firmly believe that an international science publisher has a special obligation to the environment, and our corporate policies consistently reflect this conviction.

We also expect our business partners – paper mills, printers, packaging manufacturers, etc. – to commit themselves to using environmentally friendly materials and production processes.

The paper in this book is made from low- or no-chlorine pulp and is acid free, in conformance with international standards for paper permanency.